SCIENCE & WONDERS VOLUME I

On the Edge of the Chasm

Amy Joy Hess

Ocean to Mountain Publishing
PO BOX 1116
Wallace, ID 83873
www.otmpub.com

This book is based on the true life experiences of the author, who has taken care to portray the persons and events therein with honest accuracy. Certain names have been changed to protect the guilty and innocent alike..

On the Edge of the Chasm
Copyright © 2017, 2025 by Amy Joy Hess
Created in the United States by Ocean to Mountain Publishing

No part of this book may be reproduced without written permission except in the case of brief quotations credited to the author. For information, address Ocean to Mountain Publishing, P.O. Box 1116, Wallace, ID 83873.

Books published by Ocean to Mountain Publishing are available at special discounts for bulk purchases in the United States by corporations, institutions, and interested individuals.

Unless otherwise specified, scripture quotations are from the New King James Version, © 1982 by Thomas Nelson, Inc. All rights reserved. Used by permission.

Photographs and art are the work of the author unless otherwise specified.

ISBN: 978-1-962532-07-5

To Randy

I married you because I loved you.

It mattered more to you
that I knew *you loved me.*

This book was primarily written during the years 2012-2013.

Table of Contents

	Introduction	1
Ch. 1:	Dr. Stillwell	5
Ch. 2:	The Geologist	9
Ch. 3:	Dr. Zenith	13
Ch. 4:	Randy	18
Ch. 5:	The Professors	24
Ch. 6:	das Opium des Volkes	28
Ch. 7:	The Height of Everest	36
Ch. 8:	The Flood Question	43
Ch. 9:	Three-Toed Horses	47
Ch. 10:	Stubborn Creationists	53
Ch. 11:	Hernias and Battle Axes	56
Ch. 12:	Hilary and the Roll Over	64
Ch. 13:	Tenacity	70
Ch. 14:	Blinders	75
Ch. 15:	Pancakes and Spatulas	79
Ch. 16:	When God Spoke to Me	89
Ch. 17:	Go Not Forth Hastily	97
Ch. 18:	The Lost Father	101
Ch. 19:	Kick You in the Tibia	105
Ch. 20:	Calling Heads and Tails	115
Ch. 21:	Mr. Davies and the Fire	119
Ch. 22:	Science and the Spiritual	124
Ch. 23:	Matthew	129
Ch. 24:	The Astrophysicist	131
Ch. 25:	Baa Baa Black Sheep	135
Ch. 26:	Dinosaur Tracks	143
Ch. 27:	The Laetoli Tracks	158
Ch. 28:	The Dream	162
Ch. 29:	Not My Business	167
Ch. 30:	Randy	169
	Appendix	179

Introduction

I sat up and blinked in the darkness. Moonlight shone pale and colorless across my bed.

Baron. I needed to pray for my brother Baron. The thought had jolted me awake. What was wrong? Why did I need to pray for Baron in the middle of the night?

I tried to feel around spiritually for the thing that needed to be prayed. Was he about to get himself killed? …No…but he was in trouble. Was he just fighting something? That was closer to it. He was struggling with something serious.

"Lord, please take care of my brother," I whispered. "Whatever is going on with Baron, please bless him and hold him and give him what he needs. Wrap Your big arms around him and protect him and give him strength to get through this struggle. Thank you for loving Baron, Lord. I know You love him."

What was wrong? Why did I have to pray?

I groped around in the blankets for my phone and let my eyes focus on its bright little screen. Good grief, it was 4:00 am. I needed to go back to sleep. I mushed my face into my pillow and rolled over, murmuring concerned supplications until I dropped back into dreamland.

At 6:45 I bounced awake and immediately thought of my brother. I grabbed my phone to call him, but then hung back up. "Darn, it's too early." It was only 3:45 in northern Idaho, three hours behind us in West Virginia. I'd have to wait. Setting my concern aside, I roused the children for school.

Several hours later, I parked my van under a tree down by the river and set the brake. I punched a few buttons on my phone and stuck it to my ear.

"Hello," Baron answered.

I didn't waste time with pleasantries. "What's going on? Are you okay?"

"What?" He sounded puzzled. "Yeah, I'm okay."

"Holy smokes. I woke up at four this morning with a major urge to pray for you. I just wanted to know what the deal was. You're okay?"

Baron's tone changed, and his voice pitched up. "Really?" He started laughing. "That's awesome!! I've been really struggling about certain things lately, and last night I said, 'God, would you please have Amy and Oxana pray for me?' So you did? That's great!"

"When? What time was that?"

"Oh… after midnight."

"After midnight? How long after midnight? Because I woke up at four with this huge thing where I had to pray for you. It scared me because it was so urgent." It didn't just worry me because Baron is my brother and I care about him. It worried me because Baron's troubles are generally the stuff of movie plots.

"I don't remember exactly, just sometime past midnight. That's awesome. Thanks. Thanks for praying for me. I really needed it."

That little moment in the spring of 2012 pleased Baron and me both. It pleased Baron, because he'd asked God to do something, and God did it immediately. Baron made a request and our Heavenly Father didn't even wait until the morning. It pleased me too, because God had been moving in me and showing me things on a consistent basis that spring semester, and it was nice to get some independent validation outside of myself and the stubborn old scientists in my life.

I'm a student of science. I love the adventure of exploring this world and seeking to understand the nuts and bolts of our existence. I also have a close relationship with the God of the Universe. Both of these things exist comfortably within me, whether or not others approve. I'm not loud about my faith. I don't throw it in other people's faces, because I find it annoying when it's done to me. I just live my life and trust God to direct my steps.

I had gone to school because I wanted to study the natural world with experts in their fields. I'd gone to learn, and I felt no responsibility for what other people wanted to believe about God. While at the university, though, I discovered something remarkable: God clearly cared about my irreligious science professors. They mattered to Him. By the time I started writing this story in 2012, God had been giving me guidance about my professor Dr. Stillwell for the better part of two years. Dr. Stillwell the atheist geologist. Dr. Stillwell my friend. God had spent many long years training me how to hear Him, how to recognize Him, and the wisdom He gave me about Dr. Stillwell proved spot-on time and again. I started writing this book because God moved in ways that I'd not expected - not at all.

I tried to fit the whole thing into one single volume, one book, but it started getting too thick. There was too much. Nobody wants to read *War and Peace* from a relatively unknown author, so I chopped the story into semesters. Each volume in this series fits conveniently with the events that took place during the five school semesters between the spring of 2010 and the spring of 2012. I give some of my back story in these volumes as well as events that took place later, and the entire series makes up the story of God's power working in my life and the lives of those around me.

I'm interested in understanding the universe with all its mysteries. As part of the adventure, I describe certain scientific questions I wrestled with during those years. I don't always have answers, because I just don't have enough data or expertise. Still, I think the process is part of the fun, and my hope is that some of you readers have ideas that never entered my consideration. You have insights. You have knowledge I don't have, and maybe together we can all work to fill in the gaps.

My relationship with Dr. Stillwell didn't end in the spring of 2012. We remained friends in the years after I graduated, and I visited him every time I had business in West Virginia. God continued to let me know when it was good to phone the geologist. Every time I got that nudge and obeyed it, Dr. Stillwell would declare, "Amy Joy! You called me at just the right time!" Twice I called him when I sensed it was a bad idea, dialing his office number simply because

I wanted to. Both times he answered in irritation, saying, "I can't talk right now."

The series of events has awed me. God worked more constantly and consistently in guiding me with Dr. Stillwell than He has with anybody else in my life - *anybody else* - including my kids and stepkids, my husband, brothers and sisters. I've decided there's a reason for it: Dr. Stillwell is a spiritual minefield, and I've needed careful guidance to make sure I avoid those mines and not damage what God has desired to do in that dear professor of mine.

The story isn't over, by the way. As this first volume goes to print here in 2017, I don't know how the saga ends. There are things I believe God has told me, and I have expectations about what the future holds, but the final events haven't taken place yet. I'm still waiting to see how the fifth volume plays out.

I hope you enjoy going on this journey as much as I've enjoyed writing about it.

Amy Joy Hess
September 10, 2017

Chapter 1
Dr. Stillwell

Dr. Stillwell had a minefield in his soul. I recognized it early on, long before he ever blew up at me. We became good friends because I dodged those mines, and not by my own so-spiffy wisdom. I'm still not sure how they got there, but I bet the process began way back - back before he'd built that brick wall around himself.

Maybe it started with the catechism class.

In 1954, Paul Denali Stillwell's good Catholic parents drove him to catechism classes every Saturday to soak up the subtler dogmas of the faith. The six-year-old had no interest in catechism, but he didn't fight about it. While the other children filed into their seats and waited for the priest to start, little Paul Stillwell simply sneaked away and hid.

"I told you about that, didn't I?" he asked me over his office desk nearly 60 years later.

I shook my head at the geology professor, pleased to hear one of his stories.

"I thought the priest was weird, and the whole thing seemed pretty strange to me," Dr. Stillwell grimaced across his desk. "So, when they dropped me off, I'd take a book around behind the church and read until it was time to leave."

So much for the subtler dogmas. After class every week, little Paul strolled back to meet his parents, silent about ditching the whole thing.

That worked for a couple of months. "Then, one week the priest visited our house for dinner."

I winced in my chair when he said that.

Dr. Stillwell placed his forearms on the desk and leaned forward. "So the priest wanted to know why they'd pulled me out of catechism class, and of course they said they hadn't. They'd been dropping me off every Saturday."

Little Paul was questioned. Where had he been going?

"I told them I thought the priest was weird. I thought the whole thing was pretty strange, and I wasn't gonna go."

I gazed at Dr. Stillwell, amazed by his reckless defiance at six-years-old. In front of the priest, mind you. I knew my professor's knack for irreverence, but I didn't realize it had anchored itself in his soul clear back in the first grade.

"Of course my dad said, 'I'm your father and you'll do what I tell you to do.'" Dr. Stillwell stared me down. Then, he slowed his words, speaking each one with deliberate coolness, "So, I told him, 'If you want to learn that stuff, then *you* go. I'm. Not. Going.'"

I shut my eyes at his six-year-old moxie. I'd never have said that to my father! Are you kidding? They'd have found body parts clear into Canada! Dad wouldn't have purposely murdered me, but I might have died as a side effect. I felt little Paul's impending doom.

"So, then I got a big whipping, and that made the priest happy." Dr. Stillwell rolled his eyes just a little.

I sat up to react, but Dr. S. hadn't finished. "The next Saturday, they took me back to the church. My father grabbed me and carried me inside, and I started screaming at the top of my lungs. I know he had an 'Oh sh–' moment, because when you have somebody who doesn't want to do something and doesn't care what you do to him, it's almost impossible to change his mind."

Dr. Stillwell gazed at me. "And I think you have that too."

I didn't hear him for a second. I was still astounded at his gutsy honesty. I was still focused on the fact that Dr. Paul Stillwell's rejection of his father's religion had started very early on. He hadn't waited until he was 18 at the University of Washington while his peers got shot up in Vietnam. He hadn't waited until he was out hunting down bryozoans in the Permian rocks of Nevada with

"Crazy Ernie" the bryozoologist. It had started clear back as an outspoken little boy who'd rather get whipped than hang out with a weird priest.

Am I like that too? Maybe. Maybe a little. I don't conform very well. Most of the time I can go with the flow, but I have a big *ohm* resistance to anyone who tells me what I have to think, along with this serious brain-to-mouth filter problem. I told the good doctor the next afternoon, "When there's something I really want, it doesn't go away. If I set my heart on it, I don't give up." I think that's what he's talking about.

Or. Maybe he'd just called me a mouthy little snot.

Either way, there we sat, Dr. Stillwell on his side of the desk and I on my side, kindred spirits and wonderfully good friends - friends whose religious and political views would have chewed off each other's faces if we dared release their collars.

This is a true story about my time as a science student in a small public university. Dr. Stillwell is a geology professor who enjoys rocky adventures. He's an atheist, and I'm a Christian, and we're great pals to this day. There's an interesting story behind our friendship, and everything I tell you in this book is accurate, jotted down in my journals with obsessive attention to detail.

Dr. Stillwell gets on my case about writing so much. "You know," he told me halfway down the Grand Canyon in 2011, "This is a fantastic view. But, you wouldn't know, Amy Joy. People who are *looking* at it can see it."

Dr. Stillwell loves me. I know he does. He only

Figure 1: Here I am, laughing into my journal at Grand Canyon. I'm not a boy. The gent with the tree on his back is Dr. Stillwell. He's not a boy either. Photo by Kyra Wood.

threatened to tie me to a chair and strap down my arms, a notebook in one hand and a pen (so close and yet so far away) in the other, because there's a part of Dr. Stillwell that's a sadist. Ha. Ha. Ha. Dr. Stillwell. You think you're so funny.

The point is that the events in this narrative are all true, and I have to share them because unbelievable things took place after I got to know Dr. Stillwell. These pages deal with the deepest questions of our hearts, the biggest issues we face as human beings. They contain some science, but this isn't a science book. It isn't really a philosophy book either, though it does touch on some philosophy. Above all else, this is my story, and I have to tell it because it's great stuff. The characters are rich and unique. The plot has laughter and tears, death and snowstorms. It has 2000-foot cliffs. It has fossils and dreams and ninjas. Oh yes. It even has ninjas.

And, most importantly, it contains some of the miracles I've witnessed in our staid, sterile, scientific world. That's what I want to tell above all.

Chapter 2
The Geologist

It's a peculiar coincidence that Dr. Stillwell was the very first professor I spoke to at the university. Registration day for transfers and readmits, July 2009, baby.

I wandered around campus, hunting for the Byrd Science building to sign up for classes. I explored the Byrd Center for Legislative Studies before discovering that the science building, an entirely different structure, existed 100 yards farther up the hill. Senator Robert Byrd had funneled significant federal funds into the state during his multi-decade tenure, and his wife even got a campus building named after her. Good heavens, the campus and town and state were inundated with financial Byrd droppings.

I finally located a geology room and sat down in a squeaky metal lab seat to wait for instructions. Topographic maps and posters of minerals garnished the walls. A large glass case filled the right front side of the lab, fat with chunks of rock that had followed Dr. Stillwell home at different points in his career. White calcite. Pink, wriggly-striped arkose. Black biotite that pulled apart in flaky layers. Sphalerite with its multitude of sparkling dark cleavage surfaces.

As I entered, a geology professor leaned over one of the room's black lab tables, studying papers through his reading glasses. Graying blond hair covered the lucky man's entire head, and I guessed him to be mid-50s.

Now, had I arrived there on time, I'd have heard a welcoming little spiel. And had I been signed up correctly, Dr. Stillwell might have possessed my name on a piece of paper. But I was late, and I was probably in the wrong room anyway, which are both standard for me. So, I missed out on the welcoming spiel, and Dr. Stillwell didn't have room for me.

I sat at the far lab table and waited with the other young people. If you've never been in a lab classroom, these heavy-duty, long, black tables are a constant. They can take acid spills and scalpel jabs, and blood cleans off them easily enough. There were a total of three long tables in this lab, each with a sink at the far end by the windows. Two of Dr. Stillwell's lab tables were lined with those squeaky metal stools smoothed by the back pockets of hundreds of students over the years. The third table, the one farthest from the door, was hidden by a variety of rocks and piles of paper, and its sink was overwhelmed by an industrial-sized rock saw.

Ooh! A rock saw!

The geology professor leaned over his papers. He barely glanced at me. "Come here," he beckoned with one finger.

I grabbed my hopeful class schedule and joined him. "Sir," I said.

Sir. A single word that I learned in Texas can mean, "Yes, sir," or "What are you talking about, sir?" or "Your head's on fire!" It all depends on the tone.

Dr. Stillwell grimaced. Apparently, he didn't like to be sirred at all.

"Have you looked through the schedule and figured out which classes you need to take?"

I nodded.

"Okay," he spoke to the group of us. "Well, there are a number of you here. I'm going to send some of you across the hall with Dr. Zenith."

What? I didn't want to go anywhere. In those three seconds of my "sir" and his grimace, I'd bonded with that there geology prof. About that moment, a 38-year-old astrophysicist strode through the door, ruining everything.

"Dr. Zenith!" Dr. S. shouted merrily. "Welcome!"

I considered this invader with disappointment. I didn't want Dr. Zenith the astrophysicist. I wanted Dr. Stillwell the geologist. Dr. Stillwell's discomfort with the religion of his fathers had begun at age six, and my relationships with these two men began that day.

Dr. Stillwell didn't know it, but he'd grow to love me as a daughter. I didn't know it, but Dr. Zenith would perpetually thwack me like a child who didn't want to go to church.

Physically, Dr. Zenith and Dr. Stillwell were the exact opposite of identical twins. Dr. Stillwell's iron blue eyes sat over a prominent nose that did a good job of holding up his reading glasses. Dr. Zenith's soft brown eyes matched a suitably-sized soft round nose.

Dr. Stillwell wore a plaid collared shirt. The white lettering on Dr. Zenith's black T-shirt tempted us with, "Come to the Dark Side, we have cookies."

Dr. Zenith had turned 38 that spring, and Dr. Stillwell had just celebrated his 61st birthday, but while Dr. Stillwell looked young for his age, the astrophysicist's hair had done an old-wise-man routine by going fuzzy white.

Despite his years, Dr. Stillwell jaunted about, spunky and fit, and could probably beat me up the side of a mountain. Dr. Zenith had eaten his share of Three Musketeers bars. Just a few. I'm sure he did kung fu on the side, and I didn't want to get into a street fight with him, but Dr. Zenith looked a lot more ... cuddly ... than Dr. S.

I might mention here that Dr. Zenith was of the African American persuasion. This was a prominent detail in Dr. Zenith's life, which made it important to the rest of us. Dr. Stillwell's pale skin spoke of northern lands where his Scotch-Irish ancestors used pickaxes in the heath, and Dr. Zenith was black. That's right. Dr. Zenith was black, and he stepped through the door that day with the air of an astrophysicist who thought the Dark Side had cookies.

"I get it," Dr. Zenith has said to me. "The black professor is the bad guy. The white professor is the good guy. I see where this is going. This is like the raven and dove thing on Noah's ark all over again."

Grins.

Oh. And I always forget to mention something important about Dr. Zenith's soft brown eyes; they went off in different directions. I am as serious as the Dog Star (astronomy joke there). The wall-eyed star lord regularly alternated which eye he was using, and the astute observer had to pay close attention to tell whether he was focused on you or the stuffed spherical cow on the other side of his lab. "Don't try to get away with anything. I have amazing peripheral vision with these eyes," he once said, and we all know it's true.

Dr. Stillwell served as the department chair at that time, but while Dr. Zenith technically had no great authority over the department, he did provide a lot of oversight with his amazing peripheral vision.

"Dr. Zenith is an old friend and colleague," Dr. Stillwell spoke cheerily that July day. "You, you and you go with him. He's been doing this for years and he'll take good care of you."

I had just been included with the "you" crowd.

This was the moment it started. This was the beginning.

"I keep thinking about the day I first met you guys," I told the non-twin doctors in Dr. Stillwell's back room two Junes later. The white letters on Dr. Zenith's black shirt read "Star Stuff." He had revved up Dr. Stillwell's ancient Mac computer and was playing 1970s LSD rock with glee. Dr. S. was setting up the good microscope to study bryozoans with me.

"I often think about that day too," Dr. Zenith said.

That surprised me. "The first day we met? You do?"

"No," he said. "No, I don't."

My relationship with Dr. Stillwell was put on hold after he sent me across the hall to Dr. Zenith's lab. It wasn't killed; it was merely refrigerated until the spring semester. On the other hand, any possible peace with Dr. Zenith got flipped on its bruised little head from the very beginning.

Chapter 3
Dr. Zenith

It was my fault, really. I made the mistake of telling Dr. Zenith that I'd been out of school for a few years, and I think the astrophysicist assumed his fellow Americans were intellectually lazy until they proved otherwise. He wasn't unkind about it; he just took it for granted, and I don't really blame him.

We left Dr. Stillwell's room and walked directly across the hall into Dr. Zenith's physics and astronomy lab. While natural light had poured onto Dr. Stillwell's table of rocks, Dr. Zenith led us into a cave I know once held windows before they were thoroughly masked with posters of nebulae. Models of the solar system sat atop the cupboards to the left of the door, and a picture of the phases of the moon hung above us as we entered. Dr. Zenith's stuffed spherical cow lounged in the back, watching, always watching, absorbing explanations of Newton's laws and kinematics equations semester after semester. The lab contained just two black lab tables and no rock saws, but balances and other tools of measurement filled the wooden cabinets above the counters on the right side of the room.

I'd already gone through the schedule and found the classes I needed that semester, something I had done year after year since the 9th grade. I handed Dr. Zenith my list to approve and then sat in another squeaky metal chair near those handy, tool-filled cabinets.

Dr. Zenith remained on his feet. Above me.

"You're a transfer student." Dr. Zenith held my list in his hands, reading it with his eyebrows furrowed. "What science classes have you taken?"

"Nothing... that really applies." I thought of my bio-intensive agriculture and aquaculture courses from years before. "I was a Religion and Philosophy major."

My father had recommended a philosophy major because I wanted to be a writer. "And if you're going to be a good writer, you need to know how to *think*." That was my dad's point of view, and it worked out well.

Dr. Zenith's brief squint said, "Woot."

The astrophysicist frowned down at me out of his left eye. "What science classes did you take in high school?"

"Physics. Biology. Chemistry. That was a number of years ago, though."

I shouldn't have been so honest. At least, I should have put it in perspective. I should have said, "Dr. Zenith, I haven't taken a science course in several years, but when I did, I was at the top of my classes in both biology and chemistry. I write articles on new discoveries in biotechnology and paleontology, and I sometimes read books on cosmology for the fun of it."

I didn't realize I'd have to make my case, though. I'd never faced dissension from professors, and - silly me - I expected the man to sign my slip of paper and send me off. Instead, Dr. Zenith stepped in the way of my educational aspirations. Plumb in the middle. Like a hunk of asteroid in my path toward the sun of self-actualization.

"Hmm..." The astrophysicist's eyebrows furrowed in wise concern. "I would recommend that you take the *100-level* introductory biology and chemistry courses instead of the 200-level ones."

It was my turn to grimace. "That's okay. I don't need to take those."

Let me explain this. Both the 100 and 200 level courses in question are introductory courses. The 100-level biology and chemistry are for non-science majors and the 200-level versions are for science majors. I wanted to take the more in-depth courses because I wanted to learn as much cool science stuff as I could. Dr. Zenith wasn't aware of my craving; I was there to gather knowledge and store it away in my hungry brain like a squirrel snatching up walnuts for the winter.

"You have to understand," Dr. Zenith pontificated. "This is not an easy school. We have high standards, and it would be better for you to take the 100-level science classes and succeed rather than take the higher-level classes and do badly."

"It's okay," I explained to him. "I love science. I'll do just fine in the 200-level classes."

"Why don't you want to take the lower-level classes?" he pressed, puzzled.

I wasn't prepared to answer him in any kind of civil or humble manner. A flood of reasons tumbled through my head, none I dared voice. I couldn't say, "Because I'm a *genius*, you difficult man." I wanted to avoid destruction, so I stared at the table and said nothing.

It turned out there were several things I didn't yet understand. Of utmost importance, I didn't understand that students rarely disagreed with Dr. Zenith. I hadn't been initiated into the physicist's vacuum universe free of friction; I was unaware of Dr. Zenith's reputation for extreme awesomeness. Students didn't tell Dr. Zenith, "No, sir. You're wrong. I'm not taking your advice." It didn't happen.

"I study science on my own," I tried to explain. It's fun for me. I like it.

Dr. Zenith harrumphed me. He rolled his eyes, one of them glancing at the far corner of the room. My quest for scientific wisdom outside the stringent order of the classroom clearly did not impress him. I could feel the seams of my emotions stretching.

"I'm here because I want to be a teacher," I tried again.

"Why do you want to be a teacher?" he interrupted.

"Because…" He wasn't giving me time to think. Why did I want to be a teacher? "Because I could communicate to them well. I can take hard ideas and make them fun for students, and our students need good teachers who make learning interesting and put the material into terms they can understand."

Dr. Zenith nodded in approval, but he didn't agree and send me on. He still wanted me to take the 100-level classes. Stubbornly.

Stubbornly wanted me to take those easier classes.

Had I known more about the cogs of the school machinery, I could have just said, "Thank you for your kind advice, Dr. Astrophysicist. I appreciate your concern," and skipped out his lab room door. I didn't know that all I needed from Dr. Zenith was a little sticker with a PIN on it to get online and sign up for whatever classes I wanted. Dr. Zenith had been talking down at me with a well-practiced melodramatic flare, and I believed that this man stood between me and a basic 200-level introductory chemistry class. I wanted to punch him.

"Dr. Zenith," I tried again. "I'm a writer. I do research, and I write for a living. Just because I've been out of school for a few years doesn't mean I'm stupid."

"I didn't say you were stupid. We just want to set you up to do well. We don't want you to get into a class and then find that it's too much for you. It's better to start off the lower classes. If you find they're too easy, then you can change to the higher-level classes."

"But, then I'd miss the first lectures, and I'd be behind."

"I simply don't want you to get into more than you can handle." Dr. Zenith looked at me out of his right eye. "It's important that you position yourself to succeed and to do well after a long time away from school."

This thing went on for several minutes. He would not waver! Frustration pulsed up into my chest and arms and head and pressed out my face, making my eyes bulge. I was an adult! I could make the decision to take a big-girl chemistry class without having to MacGyver a secret portal around Dr. Zenith's disapproval.

I forced myself to relax. I knew he had excellent intentions. I knew he only wanted to protect me and the school from my overconfidence. I glanced around the classroom, and the other students avoided eye contact with me.

I kept my voice calm. "Dr. Zenith, I know you don't know me, but *I* know me, and I know I'll do just fine in the upper-level classes."

That finally softened him. He asked, "What kind of math background do you have?"

"I took Calculus in high school. I got a 4 on the AP exam."

(See? If I were a genius, I'd have gotten a 5.)

That seemed to move him. Finally, *finally* he set down my registration sheet and affixed the coveted PIN sticker to it.

I don't remember exactly what he said after that. It was probably something helpful, like, "If you have any questions or difficulties, feel free to talk to me."

I do remember that I walked out of Dr. Zenith's physics lab and vowed, "I will never never NEVER take a class with that impossible man!"

Oh hoh ha ha.

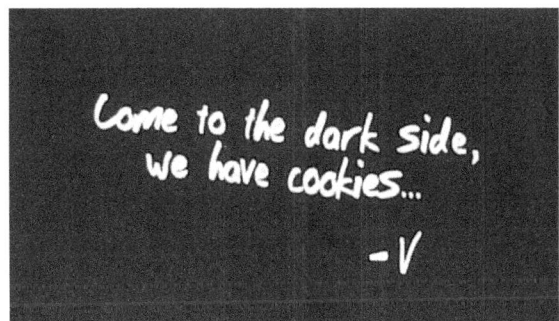

Chapter 4
Randy

It's strange to think that Randy was alive when I met Dr. Stillwell and Dr. Zenith. He was still burning glucose that day I refused to succumb to Dr. Zenith's good intentions. By the time I took classes from either professor, my husband had been gone for months.

I missed out on Dr. Stillwell's Physical Geology class that fall. It was filled by the time I registered. It started at 8:10 am, anyway, and I wouldn't have had time to get my little ones to school and still make it to class on time. Three-year-old Zeke stayed home with Daddy in the mornings, and I needed to get back to him by the early afternoon so Randy could go to work. I had a narrow school-time window.

Just to spite Dr. Zenith, I did well in my classes. On Monday, September 14th, 2009, one month into the fall semester, chemistry professor Dr. Dan handed me back a quiz with a big old 100% at the top. Eat that, Dr. Zenith!

That was the day Randy died.

We didn't get a chance to say goodbye. I was excited about the 100% on the quiz and wanted to boast to him. Sammy had bought a new Transformer as a late seventh birthday present, and he wanted to show his dad. We arrived home with a bunch of pictures of the kids I'd developed. Sammy and Savvy ran into the house to see their father, and they just thought he was asleep.

Randy had been to the doctor that morning; we thought his pneumonia might be coming back. My grown stepson Brandon and I put Randy to bed at about 1:00 pm, and he'd been snoring away 90 minutes later when I left to get the kids from school. I'd purposely stayed away to give him time to sleep uninterrupted.

I set groceries on the counter and started to open the box that held my new cell phone. Then I stopped. I figured I should check on my sick husband, so I walked around the corner, and there he lay on the bed, gray and lifeless.

"Randy!" I grabbed and shook him. When he didn't wake up, I slapped his pale face a couple of times. I put my head to his still-warm chest, listening for that familiar strong thudding. A cross-country runner in his high school years, Randy's heart had beat a solid 55-60 beats-per-minute into his middle age. Now, I couldn't hear anything. "Brandon!" I shouted. "Call 911!"

I placed my fingers on my husband's cold wrist and then on his neck. I... I couldn't feel a heartbeat at all, so I began CPR on him. That might sound horrible, but there's something immensely comforting about working to save somebody you love. Randy still felt like a human being. I was able to touch him and hold him and put my face to his face. His chest expanded as I filled his lungs, and sour air pushed back out. I breathed for him again and started pumping. Color returned to his cheeks.

"If I keep going, it'll be okay. The paramedics will get here and take over. I just have to keep it up and not stop." In the movies they always do CPR for about 15 seconds, and then somebody invariably says, "He's gone, John. Let him go." In real life, you can continue CPR for as long as it's necessary. It doesn't always work, but it can't hurt the otherwise dead person.

So, I breathed and pumped, breathed and pumped, hopeful that my husband would wake back up, that we'd get him to the hospital, that he'd be all right. Was it two breaths and 20 pumps or 30 pumps? I figured it mattered that I got oxygen into him. The paramedics would arrive and do their magic, and I just needed to keep pumping his blood through his body.

Randy had survived double pneumonia in May. He was strong. He'd escaped death in countless car accidents and motorcycle accidents. He'd been hit by a front-end loader, and he'd gotten knocked 20 feet by a backhoe bucket and kept his head on his shoulders. He'd been bitten by a feral cat in April and had to endure rabies shots, for crying out loud. He'd had shotguns stuck in his chest. He'd put up with marriage to me for nine years.

I didn't expect him to really be dead. He wasn't allowed to die.

Yet, even in the middle of the CPR, the burden lifted. Even while I pumped oxygenated blood through his body, the grief from his death already lifted up off me. I felt a relief. I felt a sense of, "It's over. He's not sick and in pain anymore." My struggling husband was gone.

When the paramedics worked on him, I took the kids outside onto the grass under the towering oaks on our four acres. I held and comforted them. I called my mother-in-law and said, "It's not looking good. You need to pray for your son, because the paramedics are here and Randy might be dead."

We didn't have long to wait. The men told me it was too late; the rigor mortis had already started to tighten the muscles in his back. Brandon told the kids to stay away. The medics told the kids to stay away. I let them keep the kids away.

I regret that more than I can tell you. If I could do things over, I'd have had the children say goodbye to their dad right then while he still looked like himself. The kids could have hugged him, and his body would have still felt warm and soft. By the end of the week when we buried him, they had him all suited up like a wax doll in the coffin. He didn't look like anybody we knew - like a mafia hit man with his hair slicked back. He was cold and hard and absolutely not anybody's husband or father or friend. Poor little Sammy went to kiss that rigid thing, that shell of what had once been his dad, and it was so wrong. I wish I'd had the little ones say goodbye back in the house, back when their father still had color in his face because I'd breathed oxygen into him.

The day of the death drained me. People kept making me do things for them, and I wanted them to all go away. The paramedics had questions. An officer arrived and made me find Randy's driver's license and describe the details of his death. I felt bone weary: I didn't want to move or think, and I sat on the front steps and stared at the yard.

"You're very calm," the officer said. "Did you take any of your husband's medications? Any of his muscle relaxers?"

What? "No!" I barked, irritated. Even that little burst of frustration felt empty; I had no energy to be angry. What did the officer want me to do, scream and wail in grief? Is that what other people did? I just wanted to sit on the steps and not move.

When the official people finally went away, Brandon approached me and asked, "Can I be the one to call Amber?"

Amber. Brandon's asking about his sister made me realize I had to tell people. I'd called Randy's parents back and told them the bad news, and that was it. I didn't think to tell anybody else. I was focused on the three little children who had lost their father, on trying to figure out how to handle this moment in which all my plans, my hopes, my heart got crushed and shoved into an ambulance with my dead husband.

I finally made myself get on the phone to contact my side of the family, but then I couldn't find anybody home. I have seven brothers and sisters, and nobody answered on the other end. Mom's phone rang and rang. I got Dad and Wendy's answering machine. Finally, my sister Whitney picked up.

"Hello?"

That's when I broke down.

"Hello?" Whitney repeated.

I couldn't talk. I tried to take in air as sobs choked in my chest. "Randy's dead," I gasped, the flood finally starting.

It's been more than three years as I write this, and I dreamt about him again last night. I asked Randy where he'd been all this time. I asked what I could do differently to be a better wife, because I wanted to fix things, I didn't like the separation between us. He smiled and said I was fine, and he played with the kids.

When I told Dr. Stillwell about the dream this afternoon, he chuckled over the phone. I know he was trying to make light of it for me. I know he didn't think it was funny.

We all returned to school as soon as possible. I wanted the

children to have some normalcy, and at school they were surrounded by supportive teachers and friends who took tender care of them. When Savannah had to draw a picture of her family in her kindergarten class, Mrs. Boggess said, "It's okay, Savvy. You can still draw your daddy in the picture if you want." So, she did.

For me, school provided a purpose. It enveloped me with people and things to do. I had always thought that if Randy died, I wouldn't be able to function, that I'd melt into a puddle in the living room and eat nothing but m&ms for a year straight. I thought I'd hate the whole world and turn bitter and enraged, that I'd roll up in a ball like a potato bug and bury myself in my hardened shell for protection. That's what I'd imagined I'd do. But, I didn't.

My fellow chemistry student Damian grinned when I dropped into the seat next to him the Monday I came back. He joked, "So, you were gone last week. Did you take a little trip to Disney World there?"

I decided to be frank. "My husband died."

"Okay. I feel like an ass."

I smiled. Dear Damian.

"Was it expected?" Chemistry professor Dr. Dan caught me in the parking lot that day. He did *not* say, "Why on earth are you at school! Your husband died last week!" He figured I'd been prepared for it.

"No," I answered. "But, he'd been in pain for a long time. He'd been in car accidents, and he'd been hit by a backhoe, and… I'm glad he's no longer in pain." The pneumonia or swine flu or whatever he'd had only added to the daily suffering that was his normal life. I was glad he wasn't in misery anymore, and it was something I could say to people, something that made sense to them.

Randy was the first man I had ever met who truly fit me, who matched me. He was the father of my children, my co-conspirator in life and best pal. He was strong and smart, loud-laughing and generous to a fault, and he loved God with all his heart. He was a rare person in this world, and I'd lost him. I should have hated life.

The thing I didn't know how to explain to Damian or Dr. Dan was that from under the pain welled a cushion of peace. The deep grieving would come, but I had an initial pad of mercy. The God who loved my husband loved me and my babies too, and despite the irreplaceable loss, I felt solid and safe. How do you tell people that?

Figure 2: Randy and Amy Joy in 2000. Photo by Whitney the Pooh.

Chapter 5
The Professors

Brittani: Is that a cheese and mayonnaise sandwich?
AJ: Yes. Don't you put mayonnaise on your cheese sandwiches?
Brittani: No.
AJ: What do you put on them?
Brittani: Cheese.
Jennifer: Turkey.

School was a huge blessing to me after Randy's death. In that refuge, I was surrounded by crowds of intelligent people who helped soothe my aching heart. Some grieving folks go hide down at the pub and some go hide in front of their computers. I had chemistry and biology and study groups. I had a host of excellent professors to liven my days and fill my mind with the wonders of the world.

Take the great biochemistry guru, Dr. Bob Manchester. Dr. Manchester was a great guy, friendly and warm and fun. He had dimples. He had that charming Lancashire British accent, and he was a big fan of the Steelers. He'd have made the perfect serial killer, because nobody would suspect him. Plus, his blue eyes bulged just the slightest bit under his mop of white hair, and if he were to drop his smile and stare you down… ohhh it would be so scary!! He'd be fantastic.

"Would you be my serial killer!" I asked Dr. Manchester one day. "I'm trying to make a movie, would you be my serial killer?"

"Sure!"

"You'd have to beat up kids," I warned him.

"Oh, I do that all the time at home."

His son Nathaniel rolled his eyes. "Whatever. No, he doesn't."

I imagine my camera on the brutal man as he slams a boy against a bedroom door, speaking his lines to the child so soft and evenly, cloaking his bulging violence with a calm facade. Then, "Cut!" and our own Dr. Bob returns all concerned and interested.

He'd say, "Was that any good? Can we do the scene again? I just don't think anybody will find it believable."

Oh they will, Dr. Bob. They will.

Dr. Bob was truly one of the neatest guys. If I'd had my druthers, I'd have chosen Dr. Bob as my mentor. Actually I *did*. I did choose him. I saw him speak on stage an hour before my first run-in with Dr. Zenith and thought, "Whatever that man teaches, I need to take it!" Dr. Bob was a marvelous combination of brilliant and fun, and majoring in biochemistry made him my advisor.

Who wouldn't love Dr. Bob? He didn't accept shoddy answers or poorly-developed research, yet he gave super good hugs and fed us chocolate truffles. He was even a good Roman Catholic, so I could talk to him about God without fear of shame and scowling. Dr. Manchester took care of us, and it cheered me to see his white hair across the food court at lunchtime.

Still. Dr. Bob is not the central character of this book. It wasn't him, it was Dr. Stillwell who became my mentor. I have a score of wonderful, smashing professors, so why him?

Dr. Dan, Chair of the Chemistry Department! Dr. Dan was dedicated, good at explaining complicated concepts, and exceptionally kind. He took time for me even when he had a pile of things to do, and he offered so much patience that he deserves an award of some sort. Or a trip to Fiji. Seriously.

Dr. Derlidger held feasts at his house for pre-med students. Beyond his background as a research biologist, he was also a fantastic cook. "Do you have any good recipes for zucchini?" I scribbled at the top of a quiz one semester. He wrote back, "Yes. I think I can find you one." He made a rule that I had to cook the recipe before he'd give me another. His entry for the chili cook-off in 2011 contained red bell peppers and orange juice and rates at the top of the flagpole of fantastic, weep-for-the-beauty-of-it chili. I couldn't duplicate its glory when I tried.

Dr. Gromp raised hissing cockroaches. She laughed freely, and I had many warm conversations with her the months I worked as her biology lab assistant. I told her about a sudden-death mammoth site in Texas; Baylor University had unearthed the bones of two dozen mammoths killed in a sudden flood. When I explained that adult mammoths had held the babies up on their tusks to keep them above the water as long as possible, Dr. Gromp's eyes sparkled, and she said, "See! I'm convinced animals have souls!"

Dr. Vallo held study sessions for us outside of class hours, helping us nail down the details of organic chemistry reactions. "What is this!" Dr. Vallo would look over our answers on his lab whiteboard. "What? Carbon has four bonds! Four. It cannot have five bonds!"

Dr. Vallo's wife and I often enjoyed tea after I helped with her housework, and she'd teach me phrases from her native Hungarian. I passed on my new words to fellow organic chemistry students to try on Dr. Vallo during study sessions. When Dr. Vallo finished helping my friend Stickley with a problem one day, I said, "Psst. Stick. Tell him '*Koszonom*.'"

"I don't understand what you just said," Stick said.

"It's Hungarian. Say it. Say it to Dr. Vallo. 'Guh-Suh-Num.'"

Stick pushed his glasses up on his nose and betrayed me. Loudly. "I'm not going to repeat the bad things you want me to say in Hungarian, and I have no problem telling on you!"

"It means, '*Thank you*,'" Brittani Love whispered to him.

"I was trying to score you some points," I said.

Dr. Vallo chuckled. "You thought you'd impress me, speaking

a little Hungarian."

Here's the point: I had a marvelous time with all my professors, and I think they're a delightful group of people. Our small public university didn't permit the professors to use student assistants. The PhDs themselves taught us and graded our papers and grilled us - without upper classmen slaves to help - and I'm sure the professors didn't get paid enough. They held degrees from M.I.T. and Penn State and Cornell, but they truly cared about teaching or I'm sure they'd have moved elsewhere. I'd serve any one of them dinner on a Sunday afternoon.

So, why Dr. Stillwell? Why did he become so important?

It really doesn't make sense. We should have been natural enemies. We each despised the worldview of the other with a passionate kind of loathing. Dr. Stillwell had convinced himself that a rational, logical person should not believe in God, and I thought he was ridiculous and intellectually dishonest. Yet a lovely friendship bloomed between us, one that seemed orchestrated by God Himself. I'm a bit mystified by it to this day.

Chapter 6
Das Opium des Volkes

"Dr. Stillwell actually struck me yesterday. Yes, he did, right in front of his wife. He hit me with a granola bar. I said, 'I can't believe you struck me, right in front of your wife.' So he hit me again." - Amy Joy's Journal

Ten of us sat in the parking lot at The Oasis in Mesquite, Nevada wondering what to do. We had to find another hotel, that's all.

Dr. Stillwell had pulled on the firmly locked doors and returned to us, drooping in body and spirit. "The Oasis is shut down," he informed us, and then he wailed, "I CAN'T BELIEVE IT!"

The pinkish hotel invited us from the huge, empty parking lot. In vain.

"You know, Dr. Stillwell," said my buddy Jared. "They have this thing called the Internet. I know you're not a big fan of technology, but if you use the Internet, you could check out these things in advance."

Poor Dr. S. had been talking about The Oasis for two days. He loved The Oasis. After sleeping on the ground for a week in Arizona and Utah, that pink hotel had promised us all the comforts of warmth and showers and good food. Lies!

"Ohhhh… The Oasis…" Dr. Stillwell moaned.

"It doesn't even look old," Katie offered. "Why did it close?"

"I know!" Dr. Stillwell said. "Those [redacted]!"

People pulled out their phones to hunt up another hotel. Not only did the Internet exist in May of 2011, but so did smart phones.

While the others investigated our options, I settled down on a curb and jotted in my journal. As I did, Dr. Stillwell's slender, silver-haired wife approached and stood over me for a moment. "You're religious," she said, obviously curious. "Don't you find that makes it difficult to be a scientist?"

This must have been bothering her. Dr. Stillwell and I had enjoyed many squabbles over politics and religion by that point, and I'd shared with him some of the miracles of my life. We'd walked miles as student and grouchy, parental professor, and I suppose his wife heard some of the feedback.

She was a sweet little woman, and Dr. Stillwell's eyes sparkled when he talked about her. Still, her question furrowed my eyebrows.

"There's no conflict with my being a scientist."

"Oh," said Mrs. Stillwell.

She didn't push the issue, and I think she should have.

Is there truly a vast chasm between people of science and people of faith? I want to deal with this ugly elephant before we go any further. Don't worry. I'll try to keep it short, but if I'm going to tell this story, then I have to answer Mrs. Stillwell's obvious question.

A passion for religion and science can absolutely coexist in one person. It can because it does. Religion and science use different methods, but they are both the search for answers. Therefore, the person who honestly hunts for religious truth can also honestly hunt for scientific fact. It's not supposed to matter which religion a scientist practices, because carbon still has four bonds, no matter who checks. *E. coli* stains gram-negative for followers of Christ and Confucius alike.

Many seekers of scientific knowledge throughout the ages have also held strong spiritual beliefs. Galileo often gave God glory in his writings. Isaac Newton and Johannes Kepler, Michael Faraday, Max Planck, and Arthur Eddington were all believers. These were brilliant men, and their contributions to scientific understanding cannot be calculated. They also came from eras when religious belief held respectability. What changed? Who decided that atheists make the best scientists, and why should any of us agree?

It's certain that colonies of skeptics had long burrowed underground, but in the 19th century the gophers of antireligious thought readily poked up their heads for all the world to see. Julius Wellhausen, Sigmund Freud, Charles Darwin all worked toward explaining the order of things without God and the supernatural, and at least some people listened.

Karl Marx famously called religion, *"das Opium des Volkes"* - commonly quoted as "the opiate of the masses." In context, he said:

> Religion is the sigh of the oppressed creature, the heart of a heartless world, and the soul of soulless conditions. It is the opium of the people. The abolition of religion as the illusory happiness of the people is the demand for their real happiness. To call on them to give up their illusions about their condition is to call on them to give up a condition that requires illusions.[1]

Marx assumed that religion existed to offer false comfort to ignorant people. He figured that if we created a fair society, a just society where people were truly happy, they wouldn't need religious comfort. He's not alone. Like Marx, the intelligentsia of our culture take for granted that the world religions developed to keep people in order, to answer the big questions of life, and to comfort widows when their husbands die. If people were educated and happy, they wouldn't cling to their nutty superstitions and the world would be a better place. That's a common view.

Julius Wellhausen declared that Moses didn't write the Pentateuch. Charles Darwin told us God was not necessary for the formation of life. Sigmund Freud decided we had created God to provide ourselves with a strong father figure. He said science and reason could now replace religion as the foundation for civilization. These ideas gained in popularity and even respectability as the decades passed.

That respectability doesn't mean they got it right, though. I was

---

1 From the Introduction of his proposed work *A Contribution to the Critique of Hegel's Philosophy of Right*, which was never finished.

taught Wellhausen's Documentary Hypothesis in my Jesuit high school, but I've since decided that it doesn't deserve the honor it's been given. I think there's better evidence that Moses wrote the first five books of the Bible after all.[2]

Certain atheists like Richard Dawkins have launched an all-out philosophical war against humankind's tendency to believe in deities. He's convinced there are no gods, and he thinks we all need to face reality. Dawkins spends the Preface of his book *The God Delusion* going through the evils that have been done in the name of one belief system or another, passionate in his war against organized religion.

It's true that people have committed unspeakable crimes throughout history under the blessing of their religious leaders. Still, the problem isn't faith in God, guys. That's short-sighted. Religious groups also set up soup kitchens and help reconstruct devastated villages and dig wells for impoverished peoples across the globe. Religion isn't the cause of the world's troubles, and it's ridiculous to say so. The problem has always been corrupt human beings. Let's repeat that, and let's use a growly voice: *corrupt human beings.* There are rotten people in the world, and they use whatever tools they have - whether politics or religion or sports or sex - to gain power and chase their desires. That only tells us one thing: humans can be awful.

I personally find religion tiresome. As far as I'm concerned, there are religious people, and then there are people who know God. In my mind, there's often a big difference between those two things. One is the facade and the other is the substance. One is the plastic wrap and the other is the pudding. In 1997, religious leader Marshall Applewhite committed suicide with 38 other members of the Heaven's Gate cult who thought they'd join a spaceship behind the Hale-Bopp comet. There are dangerous belief-systems out there. On the other hand, the self-sacrificing lives of people like Mother Theresa and Amy Carmichael demonstrate the heart of God. Jesus said, *"By this all will know that you are My disciples, if you have love for one another."* [3] Some people are clearly filled with the love and

---

2   See "Who Wrote Genesis?" in the Appendix.
3   John 13:35

Spirit of God, and some are not.

Here's what we know: whether genuine or faulty, religion is a fundamental part of the human condition. That's one reason it's a useful tool for power-hungry monsters: it comes naturally to us. All the cultures of the world embrace some spiritual belief system, and if you take the reins of the local religion, you can take the reins of the people. Anthropologists recognize that animals don't bury their dead in religious ceremonies, *humans* do. Marxist governments have failed to satisfy the spiritual needs of their people, and after seven decades of communist rule, underground churches are flourishing in China.[4] Even Europeans with their lax interest in traditional religion have embraced spiritualism and New Age occultism.[5]

If Communism had successfully created justice and prosperous societies, then maybe we could agree with Dawkins that religion is an unnecessary societal construct. Instead, communist leaders have done a good job of slaughtering millions of their own people while violently promoting atheism. Technically, atheism is a religious belief too, but you get the point. Stalin, Pol Pot, the Kims of North Korea, Mao all fought to limit religion as they built their brutal dystopias on the ideas of Karl Marx, and the result - over and over - has been the opposite of justice and prosperity. Marxism promises beauty and produces brokenness and death. Sorry Karl. Karlie. Charlie.

As atheist Carl Sagan points out, interest in the occult exploded in Russia after the fall of the Soviet Union:

> Under Communism, both religion and pseudoscience were systematically suppressed...Critical thinking - except by scientists in hermetically sealed compartments of knowledge - was recognized as dangerous, was not taught in the schools, and was punished where expressed. As a result, post-Communism, many Russians view science with suspicion... The region is now awash in UFOs, poltergeists, faith healers, quack medicines, magic waters, and old-time superstition...[6]

---

4 Rauhala, E. (2014, October 21). Risen Again: China's Underground Churches. *Time*.
5 Doering-Manteuffel, S. (2011). Survival of Occult Practices and Ideas in Modern Common Sense. *Public Understanding of Science*, 20(3): 292-302.
6 Sagan, C., (1996). *The Demon-Haunted World* (p. 21). New York: Random House.

das Opium des Volkes

Sagan blames the rampant spread of occultism on the Soviets' failure to teach critical thinking, and I won't argue with him. The Russians embraced the occult without being skeptical enough, without taking time to test the meat for parasites before they gorged. However, notice something: the Russians were still interested in sacred things after the U.S.S.R. fell. It's clear that decades of Communism didn't purge spiritual desire from the hearts of its victims. If religion were merely a cultural invention designed to keep people in order, then the Russians should have been freed from it. They shouldn't have cared. But, that's not what we saw. We saw that the Russians were spiritually starved after decades of communist oppression, and freedom gave them the opportunity to stuff themselves with whatever religious fare came available, even the horrible junk food of the occult.

Marx was premature and even naïve about the real importance of spirituality to his fellow humans. Even when every physical need is fulfilled, we find it's not enough. When people reject the traditional churches and synagogues and temples, they tend to accept some system of belief, however bizarre. It's foolish to dismiss religion as merely a social construct; it's the expression of what appears to be a universal human need.

So, how do we manage to feed our spiritual selves and remain good scientists? That's a good question to explore, and it begins with keeping ourselves honest and seeking the true answer to every question. That's where it starts for all of us, no matter our religious perspective.

How did the universe form? Where did the earth come from? What about the fossil record? What are the puzzle pieces, and what does the full picture look like?

As I entered Dr. Stillwell's geology lab, that's what I wanted to know.

Spring 2010

Chapter 7
The Height of Everest

"In historical geology, we mostly tell a story. We don't make it up that much." - Dr. Stillwell

Dr. Stillwell's class had nothing to do with my chemistry major. I could use biochemistry to cure horrible diseases or become a science teacher. But geology... I considered geology my "bad habit." I took Historical Geology during the 2010 spring semester only because I wanted to.

I've loved geology since childhood. Rocks, of all things - the study of cold, lifeless stones. I think it's a strange passion, because they're just dumb rocks. Still, those cold, lifeless, dumb rocks hold a mystery story, a puzzle that has to be worked out through clues scattered across the world. It's the whodunit of Earth's birth and life; a history book written by limestone and granite and schist.

I also learned that Dr. Stillwell grew up in Washington State. What! He came from my part of the globe!

I spent my childhood at the foot of the Cascade Mountains in Washington, and my husband Randy hailed from the Appalachians of West Virginia. We'd met and married in Idaho, but after a few years of marriage I broke it to my mother we were moving back east for a little bit.

"I told you!" Mom protested. "I told you he would move you back there!"

"It'll be just for a few months!"

A few months. Eight years. Whatever.

It turned out that Dr. Stillwell had been raised on the Olympic

Peninsula, a ferry ride from the Seattle area where I'd spent my early years. We'd both grown up with Mount Rainier in our background, and while I didn't remember Mount St. Helens before its famous 1980 explosion like he did, I felt a connection to the volcano just the same. Dr. Stillwell had moved ahead of me from Seattle to the Spokane side of the state, and we'd both spent time over the border in northern Idaho.

"When I was a kid, my friend Carla's parents took me crabbing in Port Angeles on the weekends," I mentioned to Dr. Stillwell once.

"Half the people I know are in Port Angeles," he responded.

It was like that.

"My baby sister's father owned Gloria's Steakhouse in Prichard," I told Dr. S. another day.

He barked, "Prichard! You blink and you pass it! I used to do a lot of geology work out there in the late 1970s."

Nobody has heard of Prichard, Idaho; there's not even a post office there. But Dr. Stillwell, geology professor on the east coast, knew of my tiny mountain town.

Common geography can do a lot to bond two people transplanted clear across the country. I think we enjoyed a common culture as well, foundational expectations and understandings that gave us a connection, whether we knew it or not. And there's something even more than that, even more than our common culture. The fact that our political and religious views are utter enemies hasn't prevented our friendship, because deep down, Dr. Stillwell and I are made out of a lot of the same stuff. It's like the same recipe of ingredients went into our personalities. He's got a few more pounds of iron flakes in his blood, but we understand each other.

I didn't know that at first, of course. Still, things were unique in that man's class from the beginning. Seriously. From the very beginning.

For instance, from the first day my intellect glowed with fantastic brilliance. I knew the answer to everything he asked. Don't get me wrong, I'm bright and honestly love to learn, but I'm a normal human being who makes mistakes. You know… just doofus, brain-gas mistakes. I guarantee you, I am not the smartest person you'll ever meet. But in Dr. Stillwell's class I became flaming glorious.

When I walked into class that first day, I sat at the front corner of one of the black lab tables. I didn't do it on purpose, but I'd positioned myself directly next to Dr. Stillwell's computer station at the front of the room. There he sat and made little digital drawings and notes on his Power Point slides as he taught. That meant I could speak in low tones and he'd hear me:

Dr. S: What is Occam's razor?
AJ: Basically...when looking for a solution, it's best to go with the simplest explanation.
Dr. S: What's the difference between a prokaryote and a eukaryote?
AJ: Eukaryotes have a nucleus and prokaryotes don't?
Dr. S: How many centimeters are in two feet?
AJ: Um, about 62.

Dr. Stillwell grabbed his meter stick and squinted in the dim light. "It's closer to 61."

One centimeter off. Not too bad for three-second mental math.

I tried to keep quiet, because it's irritating when some chick up front keeps spouting off. It really is, and I know it, but I'm fairly ADHD and pay attention best when I'm interacting. Otherwise, I'm planning my landscaping or worrying about my truck transmission or thinking about how many dimensions there are in the universe. As a child I used to play mute like everybody else, but adulthood has done something to me; I've lost a very healthy fear of looking stupid.

As a matter of fact, the polar opposite phenomenon took place in Dr. Bob Manchester's biochemistry class two years later. Under Dr. Bob's expectant gaze, I could not answer a question correctly for all the coal in West Virginia. It didn't matter how much I studied and came to class prepared. It didn't matter that I'd tell myself before class, "Just shut up. Just shut your mouth!" I'd still say ridiculously stupid things on a regular basis, without redemption. It got so bad that my chemistry friends started laughing at me behind my back. I gave wrong answers to questions we *all knew*, the biochemistry equivalent of looking at an A and calling it a "B." Surrounded by some of the most intelligent students in the school, mind you. I just couldn't see or hear properly. It was awful:

Dr. Bob:	Which elements have only one electron in their S orbital?
AJ:	Hydrogen and Helium.
Dr. Bob:	No. Would somebody hit her?

I *heard,* "Which elements only have an S orbital?" What he asked was slightly different. He was asking about Group I metals, like lithium, sodium, and potassium, which each have one valence electron in their outer electron shell.

Dr. Bob:	If you have 26 protons and 26 electrons, and you take away an electron, how many protons do you have?
AJ:	25.
Dr. Bob:	No!
AJ:	26!
Dr. Bob:	Okay, and if you have 25 electrons and you take away an electron, how many protons do you have?
AJ:	24.
Dr. Bob:	No! Would somebody hit her?

They did it too. Somebody would hit me. Not hard.

Top students Joe and Jessica sat behind me in Dr. Bob's class, both so calm, so smart, so quiet. Joe would refrain from offering an answer unless a professor's question dangled in the air like a swollen corpse. Joe spoke only when that body had to be cut down. Boom. Brilliant.

Not me. I longed to vindicate myself and prove I wasn't an utter boob, but I'd flop every time. It would be hilarious if it weren't so God-awful humiliating. I mentioned my stream of dumb responses at the beginning of Developmental Biology one day, and my friend Katz's eyes widened.

"Oh," she said. "You are. You're so bad. It's embarrassing."

It was.

And as I've said, I enjoyed Dr. Manchester! We were pals before I even had a chance to take his awesome classes - to shame myself daily in his awesome biochemistry classes - surrounded by the brightest and best science students.[1]

For whatever reason, Dr. Stillwell propanes the dazzling luminescence otherwise protected in my thick skull, and Dr. Manchester's presence draws out every blithering weakness I've ever had. Dr. Stillwell shines me up. Dr. Manchester chisels off my crap. They're a team.

Actually, the first wrong answer I gave in Dr. Stillwell's Historical Geology still made me look like a doggone encyclopedia.

"So, how tall is Mount Everest?" Dr. S. asked on March 10, 2010.

"About 28,000 feet."

"Well," he disagreed. "Less than 30,000 feet anyway."

After class I said, with attitude, "Dr. Stillwell. Mount Everest is 28,028 feet tall." (Excuse me, Herr Doktor Professor. How dare you contradict me!)

"Are you sure!"

"Yessir."

"Well, we'll just look it up!"

"Good idea!"

So he hopped on the Internet and googled it.

Mount Everest is 29,029 feet high. I was one digit off. One digit off… twice.

I laughed out loud as disgrace squished into every corner of my insides. I spontaneously shook Dr. Stillwell's hand and told him, "You should smack me." (Dr. Manchester would!)

"Oh, I wouldn't do that," he responded gently.

---

[1] Later, I asked Dr. Bob to write me a letter of recommendation. I really needed one from him but felt foolish even asking. He surprised me, though. He immediately stood up straighter and said, "Yes!" with confidence and conviction. Somehow… somehow… that excellent, judicious man saw past my faults and determined that I was a good researcher.

I wasn't ashamed of the mistake. I mean, who knows the exact height of Mt. Everest?[2] If I'd been gracious, it would have been perfectly okay; it was the fact that I'd had *attitude* that sent me into stammering escape-mode. Attitude is only tolerable if you're Mohammed Ali and can take on the Gorilla in Manila. If you screw up, then *at-ti-tude* is straight up mortifying.

Dr. Stillwell saw my distress and tried to comfort me. "You were close."

"No. I... the...it." I took a breath. "Never mind."

He chuckled behind me as I whisked myself out of the lab.

My memory around Dr. Stillwell has not been limited to his classroom. Things I'd never remember at other times popped into my head around him. Shortly after starting this manuscript, I said to him, "You won the Osgood Prize for your dissertation, didn't you?"

"Yes, I did. It was a good dissertation."

The Osgood Prize.

If you'd asked me to tell you what the man had won for his dissertation on fenestrate bryozoans (good grief), I'd have said, "Oh... he did win something, didn't he?" But, there in his office, the name of the honor flew from some filing cabinet deep in my brain and out my generally stuttering mouth. If I ever go on Jeopardy, Dr. Stillwell has to be in the audience. Or better yet, he has to be one of the other contestants. Close proximity to him boosts my mental faculties like jet fuel. I wonder what brand of shampoo he uses.

I'm making a big deal about this because it *is* a big deal. I have otherwise *never* been allowed to maintain that level of all-knowing with anybody. I always screw up. Any infant mouse of arrogance that creeps into my soul gets hacked to pieces with pitchforks before it can mature, but with Dr. Stillwell, I had a pass on mandated humility for one simple class of the day. It actually scared me a little.

There was another remarkable thing about Dr. Stillwell; I never bawled in his class. I'd write in my journal and get weepy in Dr. Dan's class, but I felt like an engaged, emotionally stable brainiac

---

2   I don't suppose any mortal really does. The "29,029 feet" is the official height, but there are occasional earthquakes in Nepal, and the Himalayas continue to rise 4-10 centimeters each year as shifting tectonic plates shove them upward.

in Dr. Stillwell's presence. I was set up to make a strong impression on that man from the beginning. Deserved or not.

Still. I answered the question about Noah's Flood the first week of class.

I did that.

Sigh.

Chapter 8
The Flood Question

Now, you know the Flood was going to get me in trouble. This was the classroom of a geologist, and this particular geologist had a thinly-veiled contempt for people who believed the Earth was 6,000 years old.

"*Very* thinly veiled," Dr. Stillwell agreed when I mentioned it.

I had gone into that geology class with an honest realization; I didn't know Earth's age. I really didn't. I trusted the Bible as the Word of God, but anything could have happened in those first few verses of Genesis, and I decided I'd suspend judgment about the whole thing. I had some background in paleontology, but my purpose for taking Dr. Stillwell's class was to learn what this geologist knew - because he absolutely knew things that I didn't. That was certain. I figured I'd wait until I had a whole lot of data, then I'd turn those data every which way to see how the pieces best fit together.

Dr. Stillwell was the one who asked the question about the Flood, though. Knowing him, I think he wanted to get a draw on who his young-earth troublemakers were going to be. I just answered it because he *asked*.

The lecture started with Abraham Gottlob Werner. Werner's basic view back in the 1770s was that all of Earth's geology was the result of an ancient ocean. There were biblical implications, because his ideas fit with the story that Noah's Flood once covered the entire globe. Dr. Stillwell treated old Abraham Gottlob politely, but he still explained where the man fell short.

"Werner was in bad health," Dr. Stillwell told us. "He had a hard time hiking around and studying the rocks for himself." Werner tried to do geology by reason and logic, all Aristotle-like, and it didn't work so well. My professor concluded, "If he'd gotten out more, he'd have seen earthquakes and erupting volcanoes continuing to shape the surface of the planet today."

Dr. Stillwell spoke as though Abraham Gottlob Werner represented all creationists, but I knew that wasn't so. The young earthers I'd known argued that volcanic eruptions and earthquakes and floods (catastrophes in general) built mountains and tore out valleys relatively quickly. They didn't necessarily believe that every bit of the world's geology resulted from the Flood of Noah. Creationists believed the Flood explained a lot. There were "billions of dead things buried in rock layers laid down by water all over the earth," as Buddy Davis used to sing, but there were obviously stratigraphic layers from before and after the Flood.

Dr. Stillwell finished explaining about Werner, and then he popped out, "So, what happened at the Flood?"

"You mean, the Flood model?" I asked.

Dr. Stillwell nodded.

"There was a ... There was a barrier holding a lot of water vapor in the sky, making the earth a big hyperbaric chamber and increasing the oxygen concentration of the air so that all the dinosaurs grew large." I pictured it all in my head. "It broke open and all that water came crashing down. Then, the fountains of the deep broke open, gushing water. And the whole earth was covered - "

"The added air pressure would have driven up the temperature, and it would have cooked all life on earth," Dr. Stillwell contradicted me. "Dr. Zenith and I laugh at some of these ideas, because they just wouldn't work. And where did all the water go afterward?"

"Well, the mountains didn't have to be as tall then as they are now, and-"

He interrupted, "What? There was no tectonic activity then?"

"Well, of course there was, but-"

He cut me off again.

I didn't hear the professor for a few minutes after that, because my mind had started chugging. When oceanic crust gets subducted

The Flood Question

back into the mantle, water gets sucked back with it, and there's a ton of water in Earth's mantle. That would take time, though. But, perhaps the original globe wasn't quite as blue as it is now, and whole civilizations remain forever drowned under a stubborn ocean that never *did* fully drain. Maybe it didn't drain all the way, not even a little bit!

I knew cultures the world over had legends of a massive flood that a few people and animals rode out on a large boat of some sort. Most Westerners know about the Flood stories in the Bible and the Epic of Gilgamesh, but tribes all the way to Hawaii have similar flood myths. The stories are often filled with local color and geography, but there's an ongoing theme of an earth-cleansing deluge that destroyed nearly all life.

When missionaries reached the Pacific Northwest, they learned Native American tribes had flood legends in their mythologies. One tribe told how "back in the early suns" great violence rose in the Yakima valley.[1] The good people made a canoe from the biggest cedar tree they could find and rode out the flood as it covered the mountains. The violent people were all killed.

The Frenchman André Thevet related a story by the native Brazilians at Cape Frio in the 16th Century, which described a flood that covered the mountains after a fountain of water shot out of the ground and spouted until the waters covered the earth.[2]

A legend of the Kariña people of Venezuela tells how Kaputano, the dweller in the heavens, warned the people to escape a great flood with animals and seeds in a canoe. The legend was recently made into a children's book.[3] Nobody disputes that global-flood legends abound in cultures across the earth.[4]

Dr. Stillwell's rapid-fire rebuttal caused all kinds of pictures to splash around inside my skull. If underground pockets of water were breaking open all over, that would indicate some major plate

---

1 Clark, E. (1953). *Indian Legends of the Pacific Northwest* (2nd ed., p. 45). Berkeley: University of California Press.
2 Deluge. (1810). In J. Wilkes (Ed.), *Encyclopaedia Londinensis* (Vol. V, p. 691). London.
3 Maggi, M., Amado, E., & Calderon, G. (2001). *The Great Canoe*. Toronto: Douglas & McIntyre.
4 The real question is, why are epic flood legends so ubiquitous? Do people groups exaggerate the importance of their local flood stories? Did a series of tsunamis ravage tribes of humanity scattered around the world? Or did a global flood destroy all but a handful of people, whose descendants carried the story with them as they moved around the world?

movement going on. Perhaps the Americas broke farther away from Africa, opening up the Atlantic and draining the floodwaters, dragging millions of tons of sediment into the Atlantic and Gulf of Mexico. We're used to slow plate movement, inches per year, but I pictured something vast in my mind.

Frankly, I didn't know that there had to be as much heat and pressure as Dr. Zenith and Dr. Stillwell thought. If oceans of water rushed from inside the wild young earth, then the amount of water above the firmament may not have been that great. It could have added enough pressure to warm things up a bit, to make tropical swamps grow in Montana, but not enough to pressure cook everything. All that hot water from inside the earth would have steamed things up… I needed to sit down and do the math.[5]

I hadn't actually read any kind of Flood geology scenarios. I didn't have a clue what a global Flood would be like, or how it would affect air and water temperatures. I didn't know if it would create sediment layers like we see. I could picture the huge swells, rushing tides, back and forth, burying some sea creatures on the ocean floor, surging and carrying others along with miles of limey seafloor onto the sandy beaches. I could picture the violent weather, the hurricanes and mudslides and brutal erosional forces of the water. However, my ability to imagine these events didn't make them so.

I wanted to say some of these things, but Dr. Stillwell had moved on. Besides, I recognized that I didn't know what a reasonable model would look like. Maybe my what-ifs defied physics and the geological evidence and any reasonable scenario. I hadn't investigated it yet. I'd taken the class to find out what Dr. Stillwell knew that I didn't, and at that point, I just got frustrated with him for cutting me off.

---

5   See "A Worldwide Flood?" in the Appendix.

CHAPTER 9
THREE-TOED HORSES

> Dr. S: "We have to take people based on their experience and expertise."
> JS: "I've always appreciated that about you."
> Dr. S: "Well there are some people with PhD's who shouldn't be teaching ... or breathing..." (1)

Dr. Stillwell brought up Noah's Flood the first week of class, and I'd screwed a blinking purple light over my head with my answer. After that, Dr. Stillwell tensed up on me. He didn't attack; he threw up his defense shields. It was annoying, because he kept reading into everything I said. I had to phrase my questions carefully so that he didn't assume I was trying to prove a young-earth view. I wasn't trying to prove anything, I just had questions - questions meant to provide myself with answers I actually wanted to have. We talked about Grand Canyon one day, and I asked, "Why do they think the basalt flows took so long to form? Why couldn't it have been a single event?"

"A single event, like what?" Dr. S. eyed me with suspicion. He didn't answer my question.

A few days later, I told Dr. Stillwell about dinosaur tracks I'd seen in Wyoming when I worked for an old paleontologist.

"I was at a dinosaur dig," I told him after class. "The rancher next door had been pulling back rock with a backhoe and uncovered this awesome trackway. There were big, mushy three-toed dinosaur tracks and little bird tracks. And you could see the bird track positives sticking out of the limestone that had been pulled up. You know, so, you could see both the track itself, the negative, and the positive cast of the bird track sticking out of the rock layer that had been peeled back above."

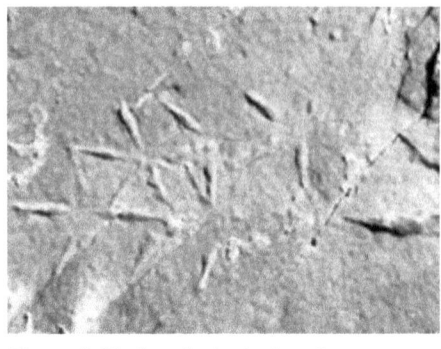

Figure 3: Bird tracks in the late Cretaceous Lance Formation on the Zerbst ranch in central Wyoming. Photo by Joe David.

Dr. S. nodded to indicate he understood me, but he refused to relax.

"The tracks must have been covered up almost instantly. You could see raindrop prints. There were these big three-toed turkey-like[1] tracks, and it looked like the animals had slipped down a slope and fallen on their rumps. There were two sets, and one of the rump marks showed the skin, but the other one had already started being swirled by water."[2]

Dr. S.' steel blue eyes focused on me. I think he was waiting for my world-wide Flood pitch, but that wasn't my point. I didn't *know* what event had covered those tracks so fast. Rain and floods and mudslides happen all the time. My point was that I'd seen this unique trackway, and I thought he'd understand how cool it was. My boss, Joe David, had walked around the dinosaur stomped-up patch of ground, taking video with his camcorder and muttering, "This is the most amazing thing I've ever seen. Wow. I've never seen anything like this. This is amazing," over and over again.

I thought Dr. Stillwell would be interested, but he didn't seem interested. He just seemed tense.

The good doctor finally relaxed a little when I brought up three-toed horses. Only another week had passed, but it seemed

---

1 I was being descriptive. *Oviraptor* tracks do look like big old turkey tracks.
2 That's what it had looked like to me at the time. The Black Hills Institute would eventually make a mold of the entire trackway, identifying T-Rex, duck-bill, *Oviraptor* and various bird tracks that Mr. Zerbst had found in a late Cretaceous layer on his Wyoming ranch. What I'd taken for slipping marks were officially determined to be tail drag marks. Casts of this trackway can be found on the BHI website: https://bhigr.com/product/zerbst-trackway-cast-replica/ . Last accessed July 14, 2024.

like eons. He'd been talking about geological strata and the fact that one location can have layers with different animal bones in them, showing that the animals living there had changed over the ages. Of course, that got me to bring up old Joe David, with his fuzzy white lamb's wool beard and squashed cowboy hat, looking like any other paleontologist you'd hope to meet.

"I used to work for a bone digger in Texas," I told Dr. Stillwell after class. "Remember that trackway I told you about? I saw that on a dig with Joe. He's not a degreed paleontologist. He's actually an artist by trade, but he's been digging up dinosaur bones for 25 years, and he's an expert at making molds and casts." I paused to explain, "They don't display the real fossils in museums most of the time. They have to display casts made from the molds of fossils. You can't put rods through r,eal *Brachiosaurus* bones and hang them from the ceiling."

Dr. Stillwell nodded in agreement. I'd been honest about the fact that Joe didn't have a degree because I didn't want to mislead my professor. To my surprise, Dr. Stillwell stuck up for old Joe.

"No, that's a man with experience, and you'll find that people with real physical experience in a field often have a lot more to offer than people with degrees who have never gotten their hands in the dirt."

Don't misunderstand his point. I've read a book or two about Schrödinger, but that doesn't make me a quantum physicist. PhDs have been immersed in their fields for years, systematically studying their subjects. They can paint each detail of the tongues and whiskers and claws of their particular discipline, while the rest of us didn't even know it had legs.

At the same time, it's foolish to underestimate

Figure 4: Three-Toed Horse (Nannippus phlegon), *from the Blanco Formation, Middle Pliocene, Crosby County, Texas.*

the man who has lived a subject every day for a lifetime, with or without the letters behind his name. Mechanics know cars. Cowboys know horses. Joe David knew dinosaurs. He didn't have the letters, but he certainly had the years.

That January day after class, Dr. Stillwell defended the value of Joe's experience. Dr. S. nodded at me, "Remember the Welsh miners we talked about? They knew which layers they'd hit in American coal mines because they had seen those exact same layers back in Wales."

That cheered me a little. "So, Joe would dig up three-toed horses and little camels and tortoises out in the cow fields. The lowest beds held mastodons, then the next layer up had the three-toed horses and camels, and the next layer had the *Bison latifrons*, which has these super huge horns that can stretch halfway across this room. And then above those we'd find the bison and cattle you see today."

Dr. Stillwell got a little excited about the three-toed horses thing. We looked them up on the Internet after class for the fun of it.

A few days later, Dr. Stillwell taught about the Solnhofen Limestone in Germany, a beautiful, thick micritic limestone in which fish and other soft, delicate fossils are preserved. Ironically, Dr. Stillwell loves to tell the story about how the Gutenberg Bible was printed using stone plates taken from the Solnhofen limestone. He adores that bit of trivia. I'm not sure why.

One of the most significant creatures found in the Solnhofen is the *Archaeopteryx*. The *Archaeopteryx*, in case you can't begin to pronounce the word, is a feathered dinosaur-like creature with a surprising mix of characteristics, including a set of teeth, claws at the end of its wings, a long tail, and theropod-like feet. If it weren't for its wings and feathers, the *Archaeopteryx* would be regarded as a little dinosaur. When *Archaeopteryx* was discovered in 1861, it was immediately seen as an intermediate between reptiles and birds. It's not that simple, though, and paleontologists are still working to understand how *Archaeopteryx* fits in.[3]

"Oh yeah," I said after class. "The *Archaeopteryx*. I've made dozens of casts of that thing."

---

3 See "What is the *Archaeopteryx*?" in the Appendix.

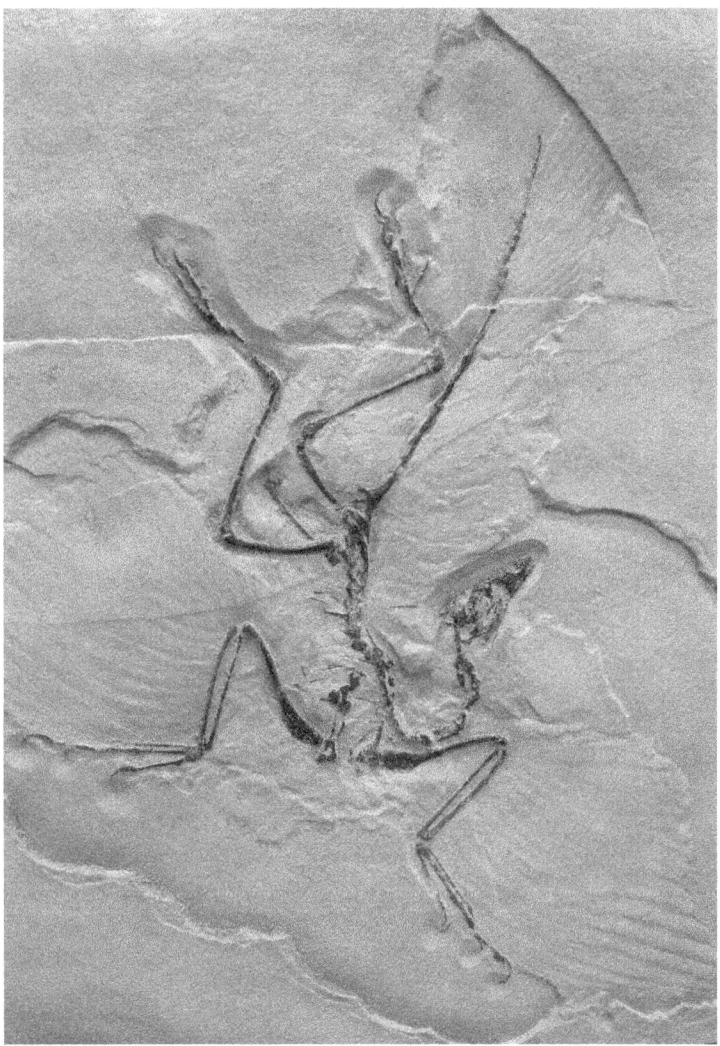

Figure 5: A cast of Archaeopteryx lithographica *made and colored by the author. Dr. Stillwell took this photograph in February of 2014. The original specimen is housed in the Museum für Naturkunde, Berlin.*

Dr. Stillwell had regarded me stiffly me for two weeks. As soon as I said those magic words, all the tension in my geology professor gushed away. It drained like turgid floodwaters through a ruptured dam. In one little statement, I'd made everything better between us.

At that point, I decided Dr. Stillwell had a dreadful lack of curiosity and made judgments based on appearances. He got all stressed out when I answered the Noah's Flood question, but he never asked me what I personally thought about the Flood. He relaxed after we talked about three-toed horses and the *Archaeopteryx*, but he didn't ask me what I thought about those either. He jumped to conclusions with hardly any information.

He's not alone. People always want to argue their points without taking time to figure out the whole story. That seems to be a common human condition. We were less than three weeks into our relationship, so I guess I shouldn't have been too hard on Dr. Stillwell.

As Joe David liked to say, "If I died and went to the bad place, I'd have to spend eternity with people who came to learn from me what I know about fossils, just to spend the whole time listening to them say what they *think* they know about fossils, which is mostly just speculation and ignorance.

Chapter 10
Stubborn Creationists

There must have been some young-earth creationists in our little geology class that spring, because Dr. Stillwell snarled and growled on occasion after our tests. He never told me who they were, but clearly someone kept insisting that the planet Earth was a young place. This aggravation had the effect of turning my dear Dr. Stillwell into Mr. Hyde the slaughter-monster. It was really an amazing transformation.

Okay, that's an exaggeration. But, a certain hostility did radiate from him.

I had expected a harmony of evolutionary thinking around me, and it surprised me to find that silent dissenters populated our very own geology classroom. I have no clue about the motivations of my fellow students. I never talked to them about it, so I don't know their reasons for rejecting the timeline of our esteemed professor.

Still, I guess I shouldn't have been amazed; Gallup polls since 1982 have consistently told us that 40% of Americans think God made man in his current form within the past 10,000 years, without any ape uncles in between.[1] Another third of Americans believe God guided evolution to create humans. There's a reason that several states had battles over teaching evolution in the late 20th century. Parents feared that schools would convince their kids to be atheists by teaching evolution. Even Eugenie Scott at the National Center for Science Education recognized this:

---
1 The Gallup (2014). Values and Beliefs Survey.

> People don't show up here (at the courtroom) because they believe evolution is bad science... They show up because they believe that if they accept evolution, then they are abandoning their religious beliefs. They see it as an either/or proposition: Either evolution happened, or God loves you.[2]

Of course, many will disagree with her, because they do believe microbes-to-man evolution is bad science, but that's another discussion. The real issue is that people believe in God, and they aren't convinced we came from apes. They aren't convinced we developed slowly from amino acids in an ancient ocean.

This is clearly a big part of the whole science-religion brawl. The general scientific community says, "This is what happened," and a variety of Americans say, "Yeah. Whatever." The scientists say, "You annoyingly ignorant religious people!" The religious folks say, "You arrogant atheists!"

The tension gets as thick as North Dakota mud, and I'm stuck in it saying, "Okay okay okay. Woah there. How can I nail down what bits are true?"

It may be true that some Americans are uneducated. They don't actually care about the science. They believe what they were told by their parents or ministers without doing real investigations of their own. They don't actually know *why* they believe what they do, but they believe it anyway. There are people who see demons around every corner, who believe every snake handler that comes along. That causes problems. But it's also not the whole story. There are others who do care about the science, who do investigate and conclude that God is very real. There are PhDs out there who love science and see problems with some evolutionary views of the universe. This is not a simple breezy matter.

I get really irritated by articles that call people "anti-science" because they hold an unpopular position. That's just cheap name calling. Bad form.

Eugenie Scott and Dr. Stillwell can complain about the failures

---

2   Ryan, J. (29 September 2005). Intelligent Design Left Dover Out. *San Francisco Chronicle*.

of science education. They can complain about the stubbornness of American religious culture. However, they may not realize that people reject microbes-to-man evolution for another reason: God still does things. It's easy to be an atheist when you see terrible things in the world, when evil always seems to win. It's easy to believe in God when you see His power working in your life and the lives of those around you.

This is a real issue, and it's one that the scientific community needs to handle well. Unfortunately, the late great Carl Sagan seems to have dismissed the possibility that actual demons do haunt our world. Not figurative "demons" or bad feelings or overactive imaginations, but rip-the-door-off-its-hinges and make the temperature drop 20 degrees literal *demons*. When they penned out their ideas about the world, I don't think Richard Dawkins or Karl Marx had a clue how much spiritual stuff still goes on in the lives of everyday people.

Today, in the 21st century, people are instantly healed in homes and cafeterias, cars and warehouses. Normal, everyday people have prophetic visions and dreams. The guy who sells paintbrushes and the lady at the deli might have encountered angels or devils. I know these people. They're my friends and neighbors and family members and random people I meet for coffee. Miraculous events still take place in the lives of people who are then stubbornly resistant to atheistic views of the universe.

It's difficult for scientists to deal with miracles when they happen. Scientists like things to be repeatable. We like to be able to run experiments under known conditions, and supernatural events don't lend themselves to scientific investigation. Still, we can't dismiss these marvels when they happen, because they do happen. We have to be ready to seek out and understand their actual causes.

I'm in a really interesting position. I'm a scientist. And, as I look back, my life has been full of miracles.

Chapter 11
Hernias and Battle Axes

My mother gave birth to me on the floor in the spare room of our house in Bellingham, Washington. Almost seven years later, my sister Whitney had the honor of entering the world on the pull-out couch in the living room. She got the hide-a-bed; I just got the floor. More precisely, I was born on blankets that covered the large hooked rug my mother had made. She says, "It just happened to be the most comfortable place in the house at that moment."

I'm proud to have popped out at home. My mother always said that her home deliveries were her best birthing experiences. Aside from Whitney, the rest of our siblings were born in the normal variety of hospitals and clinics.

The day after I emerged, my weight on the grocery store scales put me at just over six pounds. I'm told my newborn body, head to toe, fit between my mother's wrist and the crook of her elbow. I don't remember that first month of my life, but I hear that I fattened up pretty quickly.

"You were just such a chunk," Mom says to me while I'm sitting on the couch right now. "All I did was nurse you for the first six months. People would say, 'What a darling little baby,' and then they'd go to pick you up and say, 'Oh golly! What the heck! Does she have boulders in her pants?'"

I was just a little thing at first. I had reddish hair that stuck straight up and an ugly hernia shoving its way out of my little newborn tummy.

Big, nasty herniated belly button.

Of course my parents took me to the doctor, where they were informed that hernia surgery would have to wait until after I turned five. My mother had to wrap cotton gauze around my middle to protect those organs trying to escape through the hole in my stomach muscles. The protrusion had a tight, stretched quality to it, and my parents worried it might pop if they didn't take care.

Mom says, "It was so awful. It was red and bulgy and about the size of a large jack ball." A jack ball. You know, the bouncy ball that kids use to play jacks.

I'm sure my grandmother prayed for me; when she died I got her Bible filled with handwritten notes on every page. Mom didn't think to pray. She and Dad assumed they'd have to keep my tummy wrapped up and safe for the next five years. Which is why Mom was surprised by Grandma's phone call a month after I was born.

My mother says, "I remember just laying you down for a nap, rushing to the ironing board to finish ironing the curtains (which nobody does anymore) while watching my soap opera. Then I got a call from Grandma Truman. I have to go to the bathroom."

Mom interrupts her story to get up and head to the loo. We can never get through a story without interruptions. She pauses halfway across the room and looks at me.

"You just *passed over* the fact that I weighed you in the scales at the store," she accuses me. "They were the kind that hang down. You put potatoes in them to figure out how much potatoes weigh? I put you in there, and boom, down it went. There you were, my little six pounds of potatoes. And you screamed constantly unless somebody was holding you."

Which, by the way, was the big miracle on that day Grandma called; Mom managed to get me to take a nap without holding me. Mom is back from the restroom, now drinking coffee. I have to remind her to finish the story.

"Oh. What? The phone call? Well, Grandma said, 'I've been watching the 700 Club, and there was a word of knowledge that God was healing a baby girl's herniated belly button. Go look at the baby.' They didn't have carry-around phones, by the way. I had to put the phone down to go check on you. I looked, and you had a normal belly button."

Mom tells me the story, using her hands for emphasis like she always does. "And it felt so normal. Here, a miracle had just happened right in front of me, and it felt like it was a normal day and a normal thing and everything was just normal. I went back and told Grandma, 'It's totally healed. It's a normal belly button.' So, I took you to the doctor the following day and told him what happened, and he said, 'Well, it's a miracle, because these things don't just disappear without surgery.'"

(I asked my father about it not too long ago, and he said, "When I left for work that day, you had a hernia, and when I came back, you didn't.")

And that's that.

"Wait. Wait," says my mother. "And the epilogue here is… how old are you?"

I'm not answering that question.

"Okay, fine. But, through the years, when I have doubts and I need reassurance, I always say, 'Can I see your belly button?'"

She does. She does do that. And it's still fixed.

"There has to be a mechanism for that," I told Dr. Stillwell in the van between Zion National Park and Bryce Canyon in 2011. "There is an actual explanation for the fact that my hernia healed. Something real happened there."

It couldn't have been the placebo effect, because I was a baby without a clue that my guts weren't supposed to be poking out through my navel. I didn't hear my grandmother's prayers. My grandmother called my mom to go look at me and not the other way around. That's a big deal!

Dr. Stillwell didn't know what to make of it. "It's part of your family's oral tradition. I don't have enough data to make an interpretation of the event." (My family's oral tradition? We're not talking six generations here! My parents are still alive. I remember Mom and Grandma talking about it in the car when I was young.)

"Yes, but it's not an isolated incident for me," I told him. I'd experienced several legitimate miracles in my life. "It has to have a real explanation. I don't know how to handle it scientifically, but I was healed, and something real caused it."

How are scientists supposed to handle miraculous events? What do we say to those things?

Thou shalt not mix science and religion!

This might be considered the first science commandment these days. If battleaxes were in fashion, hacking sounds would have serenaded the whole science/religion brawl this past century.

Famous paleontologist Stephen Jay Gould doubted God's existence, but he tried to calm the violence by separating science and religion and forcing them onto their own plots of land. He gave them mutually exclusive domains - Non-Overlapping Magisteria (NOMA) he called them - and he decreed that one should not trespass into the territory of the other, for the good of all.[1]

Science, Gould said, determines the IS of the universe, the material facts thereof, and religion/philosophy are in charge of the OUGHT, loosely borrowing terms used by British empiricist David Hume. Science reigns over data and evidence and possibility, while religion deals with morality and ethics. Religion/philosophy work out whether human beings *should* be cloned, for instance, and science handles whether human beings *can* be cloned. Both hold important, but clearly separate positions of authority.

In the Preamble of Gould's 1999 book *Rock of Ages*, he states:

> Science tries to document the factual character of the natural world, and to develop theories that coordinate and explain these facts. Religion, on the other hand, operates in the equally important, but utterly different, realm of human purposes, meanings, and values - subjects that the factual domain of science might illuminate, but can never resolve.[2]

---

1   Gould, S.J. (March 1997). Nonoverlapping Magisteria. *Natural History* 106:16-22.
2   Gould, S.J. (1999). *Rock of Ages: Science and Religion in the Fullness of Life.* Ballantine, New York.

Gould clarifies that the two disciplines don't necessarily handle different topics; they interlace without mixing. That is, both can deal with big issues like world hunger, pollution, or cancer, but they approach them from different jurisdictions. Gould argues that scientists can derive moral codes from their religions and use them as sources of comfort, but the honest scientist absolutely ought not allow the gravy of faith to mix with the peas of science on his worldview dinner plate.

Gould didn't come up with these ideas on his own, of course. He codified principles that had been around for awhile, and his ideas have been generally embraced by the scientific community.

As a young adult, I respected Stephen Jay Gould for recognizing that a scientist could have religious faith and still be a good scientist. Gould called himself a "Jewish agnostic" but he didn't openly despise religious belief.[3] I appreciated this so much that I planned to earn a geology degree, then apply for grad school at Harvard to study paleontology under Gould himself. Never mind that every effort I made to return to school back then was thwarted and never mind that Gould died before I got my chance.

At the same time, I never fully agreed that religion and science are Non-Overlapping Magisteria. There are clearly borderlands where they fight for the same ground. Their domains touch and mush together in places, and I'll tell you exactly why. Religion is not simply about OUGHT. There are plenty of places where religion claims to deal with what IS, and when science's version of IS appears to conflict with religion's version of IS, there's tension.

Gould says that religion must not be allowed to deal with IS. If there's a conflict between religion and science about what IS, then science has dominion, because science deals with facts and scientific method and repeatability.

Rock on. That's legitimate. Facts should rule, right?

I'm going to make the case that the real conflict is not between facts and faith. Not at all. Any religious belief that is based on truth should simply fill in the gaps between observable facts.

---

3   Gould, S.J. (March 1997). Nonoverlapping Magisteria. *Natural History* 106:16-22.

Hear me out. If religion and science are fighting for the same ground, we need to start with the basic understanding that whatever is true is what's true. Earth, life, the human race have a historical, real origin. "Where did we come from?" The more good data, the more solid evidence we have, the more those things point to what's true. If we have enough data, we can find the answers to all kinds of questions, and I think God is all about our candidly hunting down the pieces of the vast cosmic puzzle. Life is all about discovery, and He can hardly be threatened by the realities of the world He invented. No honest investigation is going to concern the God of Truth.

-He who answers a matter before he hears it, It is folly and shame to him...
-The heart of the prudent acquires knowledge, And the ear of the wise seeks knowledge...
-The first one to plead his cause seems right, Until his neighbor comes and examines him.

Proverbs 18:13,15,17

We can get good answers when we have enough information. Microscopes show us bacteria and other pathogens as culprits in certain illnesses. Doppler radar, satellite imagery, super computers allow us to predict the weather with increasing accuracy. Petrified animals and plants tell us about paleoenvironments in the distant past.

When do we get into trouble? When we *lack* information. It's when we don't have enough puzzle pieces to give us an accurate picture, that's when all of us - believers and unbelievers alike - tend to fill in the holes from our own particular worldviews. That's where the fighting is, not in areas where we have solid data.

The Bible does offer a ton of scientifically keen information. The rules and regulations in Exodus and Leviticus provide real medical wisdom. Father of Modern Oceanography and Naval Meteorology, Matthew Fontaine Maury, famously studied ocean currents because Psalm 8 talks about the "paths of the seas." We learn in Job 26:7 that

God *"hangs the earth upon nothing."*

At the same time, the Scriptures were not written to systematically please scientists. It isn't any good for people to insist that the sun circles the Earth when the opposite is true. Even today, we say that "the sun rises in the east," when we know exceptionally well that the Earth moves around the sun.[4] Psalm 19 is worship, not a handbook for NASA. We have to be careful about drawing conclusions from the poetic language of the Bible, as though Heaven has literal pillars[5] or God keeps snow and hail in literal storehouses.[6] Metaphors and imagery are part of the Bible's literary excellence. We should take the Bible literally when it's meant to be literal, and we should enjoy the poetry that's intended as poetry.

The Bible demonstrates real wisdom, and God does answer a lot of our questions, but as people of faith, we should be exceedingly careful. There is plenty of room for us to go out and explore the world for ourselves.

Okay, that's one point. But here's another: science doesn't have access to all the data it requires. This is a huge problem for science, and we can't ignore it.

In geology, in biology, in cosmology there are vast gaps of information that frustrate dedicated scientists. How many historians of the ancient world would love to go back in time and watch, to have an eye-witness view! Scientists are often given the role of historians. We piece together our stories of the world as best we can from rocks and bones and DNA, and yet the mysteries abound. We use the data we have to build models, but we need so many more pieces than we have! And because we scientists all have religions, we have a tendency to fill in those gaps according to our personal worldviews. That, that right there, is where every scientist must be exceedingly careful.

In our culture, atheists are allowed to fill in the gaps from their

---

4 For example, see Alter, D. & Cleminshaw, C (1948). The Sun's Daily Path Across the Sky. *Pictorial Astronomy*, Griffith Observatory, Los Angeles, Ch. 9. We do understand the Sun flies through the heavens along with the other stars in the Milky Way. The entire universe is in motion. The Sun simply doesn't swing around Earth every day.
5 Job 26:11
6 Job 38:22

worldview, and that's wrong. Atheism is a religious view, remember. Atheists don't know whether God is there or not, and assuming an atheistic worldview does not guarantee the best answers to big questions. It just doesn't.

How did life arise on Earth? Did humans and apes evolve from common ancestors? Why does Uranus rotate on its side?

Whatever questions we're asking, atheists and Christians, Jews, Muslims, and Buddhists must all keep themselves honest. The planet Earth has an actual history. We should suspend judgment about what the data say until they create a solid picture. When there are gaps in the evidence, we can look to our models, but atheism is no more scientifically pure than any other religious position. If atheists want to keep science and religion separate, then they have to leave their philosophical views out of it too.

Hernias don't vanish under normal circumstances. The healing of my hernia had a real cause. In an interesting bit of coincidence, Dr. Jeff "Flash" Gurden told me he was also born with a herniated belly button. His didn't disappear, though; he had to have surgery to fix it when he was five.

I had a hernia. I was healed during a nap. Grandma called my mom to tell her and not the other way around. If we employ Occam's razor, I think the simplest explanation is that God healed me and used the 700 Club to tell my grandmother. If someone has an alternate explanation that fits the circumstances, have at it.

Dr. Stillwell does let me tell him stories. I'll be relating some event in my life, and I'll pause to ask, "Can I tell you the God part of this story?" Dr. S. always nods calmly. He generally responds by saying, "Well, it works for you."

Yes. But why does it work for me?

He also likes to say, "You believe what you were taught."

He should know better than that by now. He knows that I don't question only him; I question everybody.

Chapter 12

Hilary and the Roll Over

Wooden-beamed ceilings rose high above us and empty pews lined both sides of the aisle. All the people were packed at the front of the church, waiting for somebody to come pray over them. I had visited this church with a friend on a Sunday evening, and she and I watched in amazement as people crumpled onto the carpet in front of the stage. A variety of people stood around, and strange noises tumbled from their mouths.

I might have been 10 or 11. I watched the process, intrigued by these people falling all over the floor. I'd never seen anything like it. The Presbyterian church my dad attended never encouraged people to gather up front for prayer and babbling. We sang hymns and listened to sermons. I looked forward to the rope swing outside, where we took turns swooping back and forth under huge cedar trees after church. I enjoyed the hymns, but I can't recall a single one of the sermons. I tried to sleep through them by snuggling up next to my father.

This evening's strange church service offered something new. Eventually my turn came, and a man stood over me, placed his hand on my head and prayed. I swayed under his hand, and an energy rose in the excited people around me. They hoped I'd fall down, but I didn't. I knew I swayed only because I stood with my eyes closed. There was no spiritual power knocking me to the ground.

The leaders encouraged us to open our mouths and let sounds come out, and their voices rose in eagerness when syllables stuttered from my mouth. The sounds I made weren't heavenly words birthed by the Spirit of God; they were nonsense. Babbling. That was it. I left the church that night unimpressed by the whole thing.

I never *have* spoken in tongues. I've never fallen down, overwhelmed by the power of God. I'm convinced these things can legitimately take place when God shows up, but I wonder how many people are hunting for a spiritual high. I don't know whether my cynicism is healthy, and plenty of people have told me I'm missing out,[1] but it's never been a big issue with me.

Eight or seven years after that night at the church, I rode home from Thanksgiving break all curled up in the back seat of Brian Reed's car. I'd grown sleepy after four hours of driving. My college roommate Hilary Roberts had joined my family for the holiday, and we took advantage of Brian's generosity to return to Seattle for school.

Actually, no. That was our first semester in college, so Hilary wouldn't be my roommate until January.

Snow swirled through the darkness as we zoomed past Ellensburg, Washington. Fat flakes landed on the windshield in smooshy splotches. The stockyards sat out there in the cold darkness, the cattle gone, the muddy ground hard and frozen. I decided to take a nap for the last leg of the trip over Snoqualmie Pass, so I unclicked my seat belt and stuffed my coat under my head. As I snuggled up nice and comfy with the back seat to myself, an old Jack Hayford sermon on expecting miracles crackled from the tape player. So ironic.

I hadn't quite fallen asleep when Brian's car began to swerve on the snowy freeway. I opened my eyes and waited. The tail-end of the car swung to the right and back to the left.

I sighed.

"We're gonna go off the road," I thought with resignation. "We're gonna go off the road and get stuck in the snow, and it'll take forever to get us out, and I won't get my sociology paper done."

Hilary had a more emotional view of the situation from the front seat. As we nosed off the freeway, Hilary shouted out, "Jesus, save us!"

I expected to slide into the soft snow, "poof." Instead, the car

---

1 In 1 Corinthians 12:7-11 & 12:29-31 the Apostle Paul notes that people receive different spiritual gifts. I've never spoken in tongues, yet the Spirit of God has been with me my whole life, and even now I feel Him burning in my chest. The spiritual gift I've always asked for was prophecy, according to Paul's admonition in 1 Corinthians 14:1-5. That's the gift that matters to me.

surprised me by taking a tumble into the median at 50 mph, crunch, smash, dropping me about the backseat like a marble. There's the window. There's the ceiling. There's the seat. Something smacked my nose. I don't remember how many times around we went, but when the car made its last, slow "bam" to a stop, we landed on all four tires.

I sat up. "Woah! I've never been in a roll-over before."

Awesome! If it weren't for the threats of dismemberment and death, I'd secretly always wanted to live through a roll-over. I'll admit it, and you can judge me all you want.

Hilary had a more pained view of the situation from the front seat. The roof above her head crushed in as we rolled, and she had slammed her head into that jutting lump of metal. She'd been saved from an instantly fatal dent in the skull because her seat was reclined, but she'd still whacked her head. Hard. She was crying as soon as we landed, whimpering up in the front seat.

Then. Then Hilary started talking crazy talk. Words I didn't know bubbled out of her.

"She must have hit her head really hard," I said to myself. That was the first thing I thought. Hilary had gone nuts.

Then I listened more carefully, and it occurred to me that she wasn't speaking gibberish. It was more like she'd started talking in Bengali or Portuguese. I'd studied several languages by that point, and it struck me. Hilary sat in that mangled car, and fully formed words from another language poured out of her mouth. I wish I'd been able to record her to translate what she said.[2]

Seconds later, two young men from Central Washington University showed up in the snow outside our windows and offered their help. We thought they'd seen the accident, but it turned out they had youthful save-people syndrome and stopped to help. Snow and ice and dark of night did not discourage them. They helped us struggle out of the crushed vehicle, just in time for a highway patrolman to arrive. A minute later, Brian, Hilary and I huddled into the back of the law man's cruiser for transportation to the hospital.

---

2   Hilary had never spoken in tongues before that day, but she has ever since. I want her to find out what language it is - because I've learned that people who genuinely speak in tongues often speak a known language.

The officer settled into the front seat, but before we went anywhere he got out his little notepad. He asked Brian, "How fast were you going?"

"About 50 miles-per-hour."

"Did you have your seatbelt on?"

"Yes."

"Did you have your seatbelt on?" he asked Hilary.

"Yes."

"And you," he asked me. "Did you have your seatbelt on?"

"No, sir," I admitted.

"What!" the officer yelped in the front seat. "You didn't have your seat belt on! You could have been gashed open and broken your neck and killed! If your head had gone through a window, you wouldn't be talking to me right now. Always always always wear your seat belt!"

I hadn't been expecting that. "Okay," I agreed.

The officer proceeded to write a ticket, and I drooped in the backseat, scribbling in my journal and awaiting my punishment. He cited Brian for driving too fast for the weather conditions. Then, he pulled out onto the freeway and drove us to the hospital. That was it. No ticket for me after all. I got off with only that charged, descriptive lecture.

Hilary had a much worse time, because she'd jarred her neck and back. She hadn't been killed, but she left Brian's car in physical misery, which didn't improve her attitude by the way.

Those generous, save-people-syndrome young men stuck by us and offered their dorm room to us for the evening. When we called our parents, I told my mother, "Hi. I'm okay. Brian's car went off the road, but everything's fine. I'm okay. These two guys here from Central Washington University are going over the mountain tonight and offered to give us a ride-"

Hilary yanked the phone from my hand and bawled into the receiver, "We did not just go off the road! We were in a roll-over!"

Well, that ruined my plans for the night. In the end, Hilary and Brian accepted our rescuers' offer of a ride back to Seattle, while I remained behind under parental orders to take a Greyhound in the morning. Thank you Hilary. And, I did *not* get my sociology paper done on time.

That was the worst of my experience, though - the sociology paper. Hilary had been the one to cry out for help when we went off the road. She'd been the one to start speaking in another language. And she was the one who suffered from real pain.

She later said she saw angels flying around and around me as the car rolled. I think about that. I had bounced around inside the car, but I hadn't been thrown from the vehicle or decapitated. I have no doubt the highway patrolman flipped out on me because he'd seen terrible things. Yet, I was no more harmed than if I'd gone on an amusement park ride.

Poor Hilary walked around to her classes with a pillow the next few days. Brian's insurance covered chiropractor treatments, and I happily got my back and neck adjusted for free. Hilary didn't trust chiropractors and refused to go to one; her head and neck had her in agony, and she could barely hold up her skull.

I teased her and joked without mercy, forcing unwilling giggles from her. "Haha owww … don't make me laugh," she'd beg. "Hahah… owwww."

Then, suddenly Hilary was fine.

Three days after the accident, Hilary bounced up to me outside the student center. "Say something to make me laugh," she said.

I stared at her in the cool December sunshine. Her faithful pal Pillow had vanished. She stood up straight for the first time in days.

"Are you healed?" I squinted at her.

"Make me laugh!"

"What happened! How did you get healed?"

"Jon Vincent prayed for me in the cafeteria!"

Jon Vincent came from Idaho, and we liked him a lot. He'd put on this deep voice and thump his chest and say, "Remember, 'wherever you go, there you are' - a great saying by Crug the Wise. It was Crug who first invented the stick, followed by the rock, two of the great tools still used in Idaho today." Jon was a fun guy.

That day, Jon sat down beside Hilary as she lay on her pillow in the cafeteria and asked if he could pray for her. He prayed, and all her pain vanished. It was that simple.

Hilary said, "I kept waiting for it to come back. I kept waiting for my neck and head to start hurting again." But they didn't.

I don't know whether angels really flew around me that night when I rolled around in the back of the car. I'm willing to believe they did, but I didn't see them. To this day, Hilary insists they were there. I don't know what spilled from Hilary's mouth after we landed. I just know the syllables sounded like words. I do know that Hilary cracked her head and injured her neck in that accident, though. I know she suffered that week, but she healed and her pain disappeared after Jon prayed for her. She was fine after that. That I do know.

Chapter 13

Tenacity

Many years later, I sat in a geology lab at 8:00 in the morning. A geology professor who rejected the supernatural stood up front, and there weren't going to be any prayers or miracles in his classroom. Not that day anyway. It was February 5th, the Friday before the huge snowstorm of 2010. We knew that the snow clouds were blowing our way, and we figured we'd be snowed into our respective homes for a few days. I sat next to Michelle Caerphilly before class started and studied for a chemistry quiz. Around me, the other students chatted about the incoming storm.

"People will probably lose power," Dr. Stillwell said. "Make sure to have plenty of drinking water on hand. And it's smart to fill up your bathtub."

"Some of my favorite times as a kid were when we lost electricity," I offered. "We'd cook food on the wood stove and gather around reading books together by candlelight. It was fun. Maybe my mom didn't think so…" I smiled.

"Your mom probably didn't enjoy it as much as you did," Dr. Stillwell agreed.

"Yeah, but she's tough. She's the sort of person - when she was 10-years-old she'd go out on her horse with a .30-30 and shoot coyotes for the bounty on their ears."

I expected doubt from my professor. It sounded crazy. What 10-year-old little girl goes out on horseback to shoot coyotes with a man's rifle? My mother, that's who. I remember sitting at our dining room table when I was five, watching her break apples in half with her big Dutch hands. Snap! She was a tough, strong woman.

Dr. Stillwell had never met my mother, but he gazed at me for a moment and said dryly, "I wouldn't suspect any of *your* family members of being tenacious."

I studied my notes for a second before peering back at him, understanding that he'd just paid me a compliment.

"I'm being sarcastic," he clarified.

"I strike you as tenacious?"

"The question is, do *you* think you're tenacious?"

I thought for a moment. "Well, yes. But… you hardly know me."

"When you've been teaching for as long as I have, you get to where you recognize certain personalities."

That whole "tenacious" thing bothered me for a long time. I was pleased he recognized it in me, but I didn't know *why* he recognized it. Dr. S. didn't know anything about my family or my life or my decisions. He didn't know I was raising three little children in an unfinished cabin on my own. I hadn't done anything in his class but answer his questions and laugh at his corny jokes.

"All I did was sit there and look at your PowerPoint screen," I said to him once.

"And that you did very well."

Okay, but that didn't make me the sort that clung to the vine on the cliff edge, refusing to drop into the jaws of wolves. He must have heard from Dr. Zenith about our registration tussle the previous summer. He must have read my file.

Nope. He insists he saw it in me there, even back at the beginning. Before I'd ever battled him on anything. Before I'd photographed and excavated and sonicated and re-photographed and dissolved and analyzed 180 bryozoan specimens in his lab. Before I insisted we fight past the aggravations to buy 10 plane tickets to Nevada while he sat and growled and stewed. (We were going to the Grand Canyon, doggone it, even if we had to take three different planes to get there!)

Dr. Stillwell did get the "tenacious" thing right. Perhaps he saw a bit of himself in that freckled female creature who sat next to his computer station and answered questions about the Flood.

The blizzard hit, and school was cancelled for everybody for a solid week. Mounds of snow left my car an unidentifiable puff in front of the house. Snow piled high past the steps and threatened to block our front door. The children and I hung out in our cabin, unable to go anywhere for a week solid while little birds found their way in through the eaves and perched on our surround sound wires. Hello little birds. Welcome in out of the cold.

It was a good thing we had flour and eggs and powdered milk on hand. Pancakes, gravy, soup. You can do a lot with powdered milk.

Randy had owned the cabin property long before I'd met him. He'd replaced the roof and gutted the house and reframed the rooms. He'd built a little loft above the bathroom, one that looked out over the living room and kitchen "great room" below. He'd made plenty of improvements to the place, but he'd never finished it. He wanted to fix the foundation first.

The drywall had never been taped or painted. Subflooring crumbled below our feet where linoleum should have been. February was fairly breezy in our house.

The children and I chased birds out of our cabin every single day, back out into the world. We sledded down the mountain left by our neighbor's snowplow. He'd kindly plowed out my long 40 yards of driveway. Six-foot-long icicles hung from the roof, nearly reaching the top of the front steps.

The next Sunday, Valentine's Day, it snowed again. I cleared my long driveway with a shovel, not a snowblower or plow. I scooped it clean down by the main road and backed my car up so it could coast down the driveway. I didn't want to get stuck slipping and sliding on the ice in the morning. That night I snuggled into bed, satisfied we'd get out and to school on time the next day.

Monday morning came, and I chased the kids into the car at 7:05 am. I buckled them into the running vehicle and, hooray, rolled down the hill without any slipping problems. We were off, haha!

Then, boom. A wall of frozen slush met me at the end of the driveway, an early morning gift from our county plow. I hit the brakes and my car slid, smushing into a two-foot mound of ice. The small lane from our house angled into the main road, and an impassable barrier blocked the last fifteen feet.

The best laid plans of mice and men, baby.

I bellowed. I cursed! Just once. Then I hiked up to my shed and pulled out a spade.

Aaaargh!

A boulder - a *boulder* - almost as tall as the spade handle blockaded my driveway, and beyond it waited two-foot deep chunks of ice out to the road. What snow plow guys think it's a great idea to block people's driveways first thing Monday morning? Who were these monsters?

I took several pictures. Then I began shoveling. This was not a job for a snow shovel. Snow shovels are designed for loose snow. This was a job for a pickaxe.

At least I was burning calories.

By 8:30 I had made a good enough path to back up and run for it - sscccrrrloosh - through the rest.

It's not like this is an isolated moment of frustration in an otherwise smooth existence. Right now as I edit this, I'm parked in the US Bank parking lot in Sprague, Washington, trying to figure out why my van keeps leaking coolant and overheating. I'd like to get home so I can clean my disaster of a house before the appraiser shows up tomorrow to take pictures, but instead, I have 145 miles to painfully turtle down the highway, waiting for the fresh coolant to kick in. Do I have a mouse stuck in my water pump or what?

At least it's a beautiful summer evening.

That's just it. This sort of thing is normal life for many of us; we only get rare breaks in the roller coaster ride. It doesn't matter whether I love God and know He loves me, each day gets to be an adventure.

Adventure. It's the story of my life. Do you see, Dr. Stillwell? Do you see why I'm forced to be tenacious?

"Uh-huh," he said when I showed him the photos of the ice boulder in my driveway, evidence that I'd not merely skipped class. "I'd have cursed more than once."

And by the way. Dr. Stillwell is tenacious too. In a maddening, completely stubborn sort of way.

Figure 6: A county plow's Monday morning gift at the end of the author's driveway.

Chapter 14

Blinders

Dr. Stillwell had slim patience for young earth creationists, and he wasn't alone. Creationists catch a lot of heat these days because they obviously approach the science of our origins with their religious beliefs booming out before them. They start with the belief that God made the world in six days and strive to demonstrate that the earth is less than 10,000 years old. They clearly interpret the data based on what they already believe to be true, which is why most members of the scientific community tend to snap and growl and detonate like little C-4 kisses when creationists arrive in the room.

However, the scientific community is filled to the brim with hypocrites. Nobody likes to be called a hypocrite, but we need to be honest about this. Lest we make the mistake of blaming only creationists of forcing their beliefs on the data, consider this:

In 1933 a number of thinkers, including the renowned educator John Dewey, signed *The Humanist Manifesto* and thus declared their collective belief that God did not exist, that the major religions resulted from the cultural development and social heritage of humanity over the ages. Like any statement of faith, *The Humanist Manifesto* contains a list of doctrines, the first of which affirms:

"Humanists regard the universe as self-existing and not created."

It is important to understand that Humanism is a religion of its own. It's a belief system. The Humanists freely recognize this fact; that's their whole point. They had intentionally created a new religion, one that embraces a moral lifestyle while rejecting the supernatural. Humanists can be very decent people, but they draw their moral codes from their own sense of right and wrong and not a religious text.[1]

The Humanists have a problem, though. They don't actually *know* whether the universe is self-existing and not created. They don't know whether Isaiah or Ezekiel or John peeked into God's throne room. They don't know whether the Dalai Lama is the embodiment of the compassion of all the Buddhas or whether demons are simply the stuff of hallucination or mental illness or bad lighting. They don't know. They *believe*.

The point isn't whether Judaism or Buddhism is compatible with the Big Bang model. The point is that the Humanists assert *a priori* that the universe was self-created. That's the foundational belief of their religion. Yet, right now in our culture, respected scientists are expected to be Humanists. Cosmologists and geologists and paleontologists are required to interpret data through a Humanist worldview. Those who dare to suggest that the universe might not have been self-created, that it might have been created by Somebody, are shouted down and called names.

This is a serious logical issue. World! World, pay attention to this. The full spread of the evidence will *point* to whichever position is truly the correct one. The Humanists do not get to insist that their fundamental worldview *is* the correct one, right at the beginning, and refuse to hear anything else. There's no objectivity there! Their position has become so ingrained in our scientific culture that it's become "self-evident," but that doesn't make it correct.

I'm not suggesting we stick God in the gaps of our knowledge. I don't believe we should say, "God did it," just because we don't understand something; we'd never learn anything new. What I mean is this: if the universe is self-created, then the data should point to that fact. However, if a Creator really did make the universe at the beginning, there should be hints. We don't have to stick God in

---

1 This isn't without fault of course, because we humans can rationalize anything as "good," which is how we got the Holocaust. However, the Apostle Paul agreed in Romans 2:14-15 that we do have an internal compass, a conscience, which shows the work of the Law written on our hearts even when we don't have it written on paper in front of us.

the gaps, but if God did make everything, the forced blinders of Humanism will prevent scientists from seeing the evidence for it.

My fellow scientists obviously understand the minute details of their specialized areas of research - details that the common reader of *Nature* has no inkling about. They spend years at it, and I expect that most of them try to be honest and as objective as they can. It's unlikely my questions are new ones. There's nothing new under the sun, after all.

But, we can question our fellow scientists too. Science is all about questioning. It's about double-checking and testing and finding holes in each other's logic. Newton's laws stood for centuries until Einstein came along, a patent clerk, for crying out loud, and pointed out the weird realities that take place near the speed of light. Newton's laws work just fine until objects approach the speed of light.

The general public has a bleached and dolled-up view of scientists, as though they are a separate species, brilliant and all-knowing, unhampered by cataracts, gazing on the world from a clear vista with no trees or wreckage to block their view. The reality is that scientists are prone to peer-pressure and short-sightedness like anybody else.

I laughed out loud when I read how the great evolutionary biologist J.B.S. Haldane used to treat speakers he didn't like. He'd sit in the front row, holding his head while his pained cries of "Oh God! Oh God!" reverberated through the room.[2] I suspect he wasn't praying, and it makes me think that old J.B.S. was a bit of an egomaniac.

Guess what. Scientists joke around and play with their dogs. They cry over sad commercials and worry about their kids. They get drunk and come to work hung over. (Alcohol flows freely at geology conventions.) Scientists can struggle with dyslexia or OCD or manic depression. They can catch sexually transmitted infections. Some cheat on their wives or eat too much junk food or get petulant and say horrible things to their sick spouses. In other words, scientists are just like other human beings.

---

2 Charlesworth, B. (2004). John Maynard Smith. *Genetics*, 168(3): 1105-1109.

No scientist truly leaves his or her most cherished beliefs at the lab door. All scientists are affected by their personal worldviews and prejudices. Some of us have seen miracles, and some of us have been whipped in front of the priest. Some of us have rejoiced when our beloved ones were healed, and some of us have suffered while our beloved ones died. There's no escaping our personal baggage.

Science can be a high pressure business. There's money involved. There's reputation and clout at stake. There's a heap load of arrogance, and (this aggravates so many of us) no one person can see the whole picture by himself. We scientists are still humans with individually limited perspectives. We must be ready for new explanations that fit all the data, explanations we've never considered.

I'm hunting for the answers to honest questions. I don't blindly trust my fellow scientists who have embraced the Humanists' worldview, because I know they've filled in the gaps from *that* perspective. They might be right and the evidence might indeed point to their conclusions, or they might have overlooked some game-changing pieces of the puzzle.

I freely admit I come at the issue of "how did we get here?" with a belief in a vast, powerful God. I don't believe in shoe-horning data where they don't fit, and I am not threatened by whatever is the true history of Earth.

What do I want? I want to understand it myself, to peel off the extra fat and skin of interpretation people love to use to flesh out the bones of evidence. I just want the bones - the ground-true facts. I am weary weary weary of opinions.

Chapter 15

Pancakes and Spatulas

There are milestones in every relationship, moments that cook two people together like running puddles of pancake batter or, you know, cut them apart like a spatula. Dr. Stillwell and I had one of those moments early on.

Even during the blizzard week, I'd spent time discussing issues with Dr. Stillwell in my head. I had questions, puzzling, driving questions about the history of the planet and where geologists got their ideas. Why did paleontologists resolutely insist that humans and dinosaurs never co-existed? Who decided that and why? Why did they think the Appalachians were once 40,000 feet tall? How long did it take things to erode? How did the Grand Canyon really form?

When school resumed after the snowstorm, and after I conquered the snow boulder, I asked Dr. S. if I could discuss my geology puzzlings some day after class. He said, "Sure!"

Then weeks passed.

Soon the snow melted, and the sun warmed the world through barren branches overhead. One particular March morning demanded outdoor lessons. Dr. Stillwell opened the lab doors and led us outside, where we felt like small children on the first day of spring. We met the ancient Dr. Bell hiking with his botany class, teaching them how to identify trees according to characteristics of their twigs. *Just their twigs.* It was that type of fresh and damp old-school-learning day.

I had some brief business to attend to and called the county probate office on our way into the sunshine. They'd been leaving me messages regarding Randy's estate, so I called and found out I needed to go in and sign some paperwork.

Kinda depressing.

I tailed behind the rest of the class as they reached the northwest corner of the library. Just as I ended the call, a gleam of light glanced across my eye. I squinted as another glint flashed in my face. Dr. Stillwell had his reflective compass aimed at me, purposely stabbing my eyeballs with light.

The other students tittered. Dr. S. looked pleased with himself. I got the hint and snapped the phone shut.

The good doctor held out his compass and pointed at the library. "Do you see how the buildings are laid out? They don't run north-south-east-west but are lined up at North 20 degrees East. Anybody have an idea why?"

We all shrugged, even big, shaggy-faced Whitaker, who usually had something entertaining to say.

Dr. Stillwell relented and told us. "Because that's the strike of the rock bed below. It was easier and much cheaper to build in the same direction as the rock layers than to try and blast across them."

Dr. S. moved on, leading us down the walking path to the football field, where he pointed out swirling circles of stromatolites in the boulders that stuck out of the grass. I didn't know that stromatolites were mounds of cyanobacteria and trapped sediment. I just saw the circles they had formed in the rock, preserved from time immemorial. Stromatolites! Right there in the rock by the football field, leftover from an ancient ocean!

I studied the boulder with interest, imagining the waters that once splashed over us. Then I slumped onto the grass next to the boulder and listened to Dr. Stillwell. I hadn't slept much the night before.

The first few months after Randy died, I felt okay. It's hard to explain, but I had tremendous peace about his death. I missed my husband and blubbered regularly in the car on my way to and from getting the kids from school, but I didn't feel destroyed. I'd been alone for years before I'd married Randy, and I knew how to be alone. I'd married Randy because I loved him, because we matched each other, because he was my friend and I didn't want him to go home somewhere else every night. I was able to survive with him gone, but as the months passed, my heart ached more and not less. He hadn't simply gone on a trip; he would not be coming back.

When Dr. Stillwell finished talking about the stromatolites and the Conococheague Formation and other things hard to pronounce, our stream of geology students headed back down the path. I trudged along, hands stuffed in the pockets of my thick, baggy sweatpants. My heart hurt, and I was tired. I wanted to go lie down somewhere and sleep.

Dr. Stillwell had a lively, fun personality, and he took personal interest in his students. He knew which students were getting married and which were having babies. But, he had a sturdy wall set up between knowing *about* us and actually getting involved. He protected his personal space and didn't let people past that wall. He may not have meant to walk around his barrier that morning, but he did.

He strolled up beside me on the damp path, the sky bright and blue through barren oak branches over our heads. "Are you doing okay?" He asked it as though he honestly wanted to know the answer.

I didn't think about it at that moment. I didn't even look at him. I nodded automatically, the way you say, "Fine" whether or not you're contemplating homicide.

He didn't accept my nod and move on. "Are you lying to me?" It wasn't really a question.

I smiled inside at that. I nodded again, slower this time, then glanced up the path behind us. Ten paces separated us and the last lagging students. I turned back to him and confided, "My husband died, I don't know if you knew that."

Dr. Stillwell's eyes widened. "When?"

"Six months ago, and I'm okay... I'm just having a day." I tried to reassure him. "I'm all right, I'm just having a day."

From September 14 to March 9. That's 176 days.

Years later I asked him, "Why did you do that? Why did you ask me how I was doing? You specifically don't want students crying in your office, so why did you do that?"

"I don't know," he said. "Something told me it would be okay."

I didn't bawl in Dr. Stillwell's office that afternoon or any other afternoon. I did get to have my first real conversation with him after lab, though. We returned to his lab classroom. The other students departed. I asked the professor if he had a few moments to talk geology, and he said he did. I hung about after class, looking forward to getting some things explained.

I should be so blessed.

Here we had an opportunity to discuss deep bubbling questions about the history of the world and feed my hungry mind. We had an opportunity to open new vistas for my understanding and set me straight and ensure I didn't have a bunch of ridiculous notions mucking up the insides of my head. We didn't get to do those things though, because Dr. Stillwell ruined everything. He decided to drop off our gorgeous geological cliff into the abyss of … politics. Yes. He left my beloved geology behind and plunged right into the sewer.

In the years Dr. Stillwell and I have known each other, we have rarely argued about religion. We've discussed faith and God and religious beliefs, and we've disagreed with each other, but we don't squabble about it much. It's too emotional a subject. We've never argued about geology, not even once. There have been many multitudes of verbal fist fights about politics, however. Politics of all things!

Why! Why stupid politics?

Because he starts it! Because he's a doggone socialist, and he starts it. He's a shameless socialist, and I am a fairly conservative sort of free enterprise American. The good doctor and I were so bad at the Grand Canyon that next summer that fellow student Kyra said, "Would you two STOP IT!"

It *irks* him that my religious and political views conflict with his. It really does. And yet, politics is the safe zone representation of our worldview differences. We can hold strong positions without exposing our deepest hurts. We can shred each other without getting personal. And Dr. Stillwell enjoys it. We have growled and snarled and snapped at each other at times, but deep inside he thinks it's fun. He likes to pick fights with me.

And me? I see politics as an abscess on the gums. We absolutely should be involved in the political process - ours is a country *by the people*, which means it's our responsibility - but politics isn't going to save us. It doesn't matter how often we pop the abscess, it's the rotten tooth that's the real issue. The face of the political monster isn't as important as what's going on inside people's hearts. It's the dark and cruddy hearts under the surface that cause the real problems of our world.

I lingered at the middle black lab table hoping to air my geology confusions. Instead, Dr. Stillwell started talking about the deterioration of society, and he used the opportunity to attack tax cuts for the rich (He actually said that). When I rejected his tired propaganda, he harangued me.

"Ohhh," I groaned. "Tax cuts for the rich is a loaded argument, because the rich pay a ridiculously high percentage in taxes."

"But they don't pay more," he countered, "because they find loopholes!"

"Well, good for them. I wouldn't mind a flat tax. We don't want to tax the wealthy so much that they hide their money offshore. We want to keep wealthy people and businesses in America."

"Then you're advocating Social Darwinism!"

"What?" I squinted. "I don't see the link."

"The rich just keep getting richer on the backs of the poor–"

"Oh…" I finally caught his meaning. He didn't like the big, cruel capitalist machine. "You mean survival of the fittest? Look, I just don't think it's right for me to take money from one person because I want to give it to another person. It's not *my* money. What right do any of us have to decide what other people should do with their own money? 'Hey! This guy over here needs shoes. You have money. I'll just take your money and buy his shoes.' That's so arrogant!"

"Their money?" Dr. Stillwell sneered. "Money they got by stealing from other people? The Republicans have this attitude that people should grab their own shovels and start to dig. 'Dig grandma, dig!'"

Dr. Stillwell has this way of making my whole face squish in bewilderment. Where did he get this stuff?

"You don't know how they got their money! And if my neighbor is in trouble, I'm totally willing to help him. But, I'd much rather give my neighbor $100 than have the government *take* my $100 in order to give my neighbor $50."[1]

Dr. Stillwell rebutted, as cheerily as could be, and I re-rebutted. He was enjoying the banter with a lively wickedness, but it got to where I didn't want to answer him anymore. He wasn't interested in honest conversation; he kept spouting leftist slogans.

"Look. I don't really care. I don't want to talk about politics, I want to talk about geology."

He grinned and kept on talking.

"I don't want to talk about politics," I insisted. "I want to talk about geology."

And with a *glint* in his eye, he kept arguing. He was having fun. "You just contradicted yourself. You want the government to fix a problem when a moment ago you said the government was the problem."

Amazing man! There's a happy medium between anarchy and invasive metastasized Big Brother Robin Hood!

I chose to change the subject. Forget him and his silly political interruption of my day. If we had two minutes to talk, I didn't want to waste it on whether I respected Steny Hoyer or despised Harry Reid.[2]

My heart hurt. It hurt physically. Every day. Day after day.

"Where does grief come from?" I said suddenly. "What causes it?"

That simple question swept away Dr. Stillwell's merriment. All the delight dropped from his face, and sober blue eyes stared back at me. That wasn't my intent, but I appreciated him for it.

"I have food and water and clothes and a place to live," I went on honestly. "I have people who love me. Why do I feel grief? What

---

1 I really do help people whenever I can. In fact, I have two additional adults and three additional children living in my house as I write, and nobody pays me rent. They help with food and chores. Dr. Stillwell does his part too. He takes in stray kitties. The world would be a lovely place to live if we just cared about each other - wretched governments notwithstanding.
2 Which I did at the time - respect Steny Hoyer and despise Harry Reid. In 2010, Hoyer and Reid, both Democrats, were respectively the Majority Leaders of the House of Representatives and the Senate. When I mentioned them both by name, Dr. Stillwell relaxed the slightest bit.

purpose does it serve?" My point was not to get him to feel bad for me. I didn't feel bad for me. I had a much more philosophical question in mind.

I was suffering. On our walk, Dr. Stillwell had shown interest in me because I was obviously in some weary pain. And for what reason? What purpose could he give from his Humanist worldview? If we were merely the result of biochemical changes over the eons, if we had evolved step by step from the kin of stromatolites in a primordial ocean, what evolutionary advantage did grief possibly have?

Grief sucks away our energy and leaves us struggling to find a reason to live. Grieving people can spend months sitting, doing nothing, staring at walls. Not eating. Not sleeping. Grief makes people want to eat chocolate and drink whiskey and do drugs to find relief. If we evolved from bacteria, we should never have bred grief into our DNA. We should have developed as eating, mating, killing machines, but we do feel grief, despite the fact that it makes us weak and less likely to survive. If we bonded with each other for the purpose of survival, then we should have a much better mechanism in place to deal with loss.

If we were made to live forever, though... if we were made to develop deep bonds of friendship and love, then grief makes sense. We weren't meant for death. We were meant for life, but we've inherited a broken world. That makes far more sense to me than any evolutionary explanation. It really does.

If grief is just a bi-product of love, then what business did *love* have in Dr. Stillwell's worldview, for that matter? People argue that love helps us to survive by way of collective social care and comfort. Elephants surround their babies to protect them, and that protective nature does preserve our young. But love, self-sacrificing love, has nothing to do with evolution. It causes us to prop up the weak and sick and broken among us. If human life has inherent value regardless of its evolutionary fitness, then it's right that we should nurture and care for the child with hemophilia, the asthmatic and the diabetic. But, that's not Darwinism. If Darwinism is our model, why are we letting those with major defects infect the collective gene pool?

Why? Because there's more to life than survival of the fittest!

I stood at Dr. Stillwell's rock-strewn lab table with my aching chest, frustrated that this man who rejected Social Darwinism could embrace a worldview where grief made no real sense.

(I'm lying on my couch with my laptop as I write this, and the tears are dripping out the corners of my eyes and down my jaw and into my ears. I'm not even sure why.)

Dr. Stillwell dropped his political boxing gloves and listened to me, genuine concern on his face.

I explained out loud, "I don't hold it against him for dying. I loved my husband. I really did. Not just for me, for *him*. I loved him for his sake, and I'm glad he's not in pain anymore."

He nodded.

"I hate that my kids lost their dad. The real estate appraiser came to appraise our cabin recently, and my brave little boy ran over and hugged this strange man's legs."

Pain creased across Dr. Stillwell's eyes.

"I said, 'Zekie, what are you doing!' He said, 'I'm giving him a hug!' I was like, 'Oh Zekie, you need your father.'"

(That little act of affection touched the poor appraiser's heart. He felt so bad, he gave us the appraisal free of charge.)

I hadn't talked to anybody about Randy's death in a long time. Nearly six months had passed, and I didn't seek out help from people. In that geology lab, I began verbally hemorrhaging. I had started down a thorny path and didn't care a bit. Anything, I guess, is better than tax cuts for the rich.

I leaned forward and informed my professor, "I would never have divorced Randy. I might have shot him in the face," I grinned, "but I wouldn't have divorced him."

Dr. Stillwell chuckled, "There is that." He'd been married many years. He knew.

"I do understand the PMS defense for homicide though. Randy's pain meds would make him loopy, but he wouldn't go to bed. He'd stay up and try to cook himself food, and all he'd do was make a mess. If he'd put his head down on the pillow, he'd be out in 20 seconds, and he desperately needed sleep. Plus, he was supposed to be watching Zeke in the morning. But instead of letting me cook for him at one a.m. and going to bed, he'd stubbornly take three hours to make himself a bowl of cereal." I'd wake up to find coffee beans in a bowl of milk, next to the stove where an unidentifiable goo had poured into the burner hole.

"I was PMSing one night, which, normally I don't. I don't normally PMS. But, this night I just had no patience left. I became completely childish and petulant and said horrible things to him. I was so furious, I wanted to shoot him in the chest. I even went to the closet for the shotgun."

I had no moral sense of the wrongness of it. Filled with aggravation, I really wanted to shoot him. Even then, I paused when I reached the closet.

"What stopped me were three things. One, the kids needed their father. Two, they needed their mom, and while I wasn't worried about going to jail-"

Dr. S. nodded, entertained.

"-I knew that if I *went* to jail, I wouldn't be there for them, and they needed their mother." I raised my eyebrows. "And number three, God didn't want me to kill my husband."

Dr. Stillwell brightened, pleased with himself. He didn't say it out loud, but his eyes said, "Haha! I was right! I knew you believed in a deity!"

Four days after that angry night, Randy never woke up again, and I thought, "Oh no. I killed my husband. I wanted to shoot him and he died!" I'd gone to the closet for the shotgun with murder in my heart, and four days later he was dead. I didn't really want him dead! I'd wanted to shoot him, but I didn't want him dead!

I met Dr. Stillwell's eyes, serious. "There are some people that you should be very glad believe in God, and I'm one of them. I would be a dangerous person if I didn't believe in God."

"I think you still are," he said. "You just keep it in check."

I looked at my professor, and he looked at me.

I missed my dear husband, and at times I ached with shame

wishing I'd been kinder to him. I think of all the ways I could have been a better wife. I had three little children who'd lost their dad, and I had to live with the daily reality of this gaping hole in my chest.

Dr. Stillwell took a breath. He didn't answer my question about grief. He didn't give me any scientific explanations for its existence in the world, but he did give me a bit of fatherly wisdom.

"You are going to be in mourning for a long long time. It doesn't just go away."

I felt my pupils dilate. Those few words enveloped me for a moment with the burden of living for "a long long time" with my torn heart.

It took some time to wiggle its way into my understanding, but it occurred to me that Dr. Stillwell had gone through some distinct pain in his own life. When he told me, "It doesn't just go away," he said it from experience. People don't love life because it's easy or because they're born with a cheerful personality. People love life by choice.

Chapter 16
When God Spoke to Me

I used to get angry at people who said that God spoke to them. They'd say, "Well God told me…" and I would say, "How! How did He talk to you? Did you hear a voice? Did you hear Him in your head?"

No. No, they said He'd spoken to their hearts.

That didn't impress me.

I'd insist, "God didn't talk to you! That's your own thoughts. Your feelings. Your ideas. You are crediting God with your own feelings and thoughts." I wasn't polite about it either. I barked and snarled much louder than necessary and, to their credit, each person patiently let me bawl out my skepticism.

Those people weren't the real cause of my frustration, anyway. I possessed a massive longing for God to talk to me and resented anybody who spoke lightly about hearing from the King of the Universe. It wasn't that I needed proof of His existence. I knew perfectly well He was there, but I hated my one-sided conversations with my Heavenly Father. I hated speaking to the sky and hearing nothing in return. I couldn't ask God questions and get straight answers. I couldn't hear His voice. I was stuck, and I hated it.

And I really needed God to be there for me. My parents separated when I was seven and my earthly father lived 350 miles away. I missed my big brother in a very deep way; he'd left home when I was nine and, except for one cold Christmas, he hadn't come back. I had six younger brothers and sisters and an overworked mother, and nobody looked after me to make sure I was okay. I can't tell you how much I longed to hear God's love for me in His voice.

About 10 p.m. one evening the summer I turned 19, I decided I'd walk nonstop until God broke His silence. I'd heard of people doing such things, hanging out in the woods until God relented and gave them some personal One-on-one direction. I knew I could keep on and on and not give up, so I started walking down our lonely country road. I decided that I'd walk the 17 miles of wilderness to the freeway and keep on going; I'd walk for days if I had to until God said, "Hello Amy Joy."

I didn't have the attitude that I could force God to do anything; I wanted to demonstrate His overwhelming importance to me, to prove I really meant it. After all, the Bible says God rewards those who hunt for Him.

And you will seek Me and find Me, when you search for Me with all your heart.

Jeremiah 29:13

…for he who comes to God must believe that He is, and that He is a rewarder of those who diligently seek Him.

Hebrews 11:6

I hiked down our long dirt driveway and onto the moonlit pavement under towering cottonwoods and firs and cedars. I listened as I walked, because people said, "If you want to hear from God, you have to listen." I walked and prayed. I walked and shouted. I walked and sobbed - miserable, lonely, frustrated tears.

The river gushed to my right. Crickets and hisses and rustles and other music of the night entered my ears. I heard noises in my head, the echoes of sound vibrations my brain had absorbed during the day. My neurons coughing. None of it was God.

I had ambled along exactly eight-and-a-half miles when God spoke to me. Halfway to the freeway. And not through my neurons. He did not talk to me out loud. He didn't whisper in my head or my ears. It was as though He spoke directly to my spirit, and I recognized Him. I knew Him. I heard Him, even though He didn't use an audible voice like I'd wanted.

When God Spoke to Me

He said, "Go home." He said it simply and firmly, and that was it.

"Noooo!" I shouted into the night air. "I want You to talk to me!"

I was only halfway to the freeway! I'd been planning to walk and walk and walk until Kingdom Come, and I'd hardly even started! I strode on several more feet in frustration.

Then I stopped. It was no good. If God tells you to go home, it's no use to keep heading on, hoping He'll open up and start chatting. Disobedience wasn't going to get me anywhere, except to the freeway. Which was lame if God wasn't going to talk.

I turned around and trudged toward home. Weary. Disheartened. I plodded the eight-and-a-half miles back up the road and reached the house by about six a.m., where my mother ordered me to bed.

"Where were you!" she said later. "I thought you were out behind the barn. About two in the morning, we prayed, 'Lord, where is she? Wherever she is, send her home.'"

I thought, "Thanks a lot, Lord. That's not what I wanted."

I didn't have a rude heart toward God. I just wanted Him badly, and He knew that. I longed for a real relationship with Him, one where I didn't do all the talking while He remained as mute as the stars overhead. I wanted to know Him and enjoy Him and spend time with Him. I knew it was a rare moment in history when God had spoken with Moses face to face, but I desperately needed more than a silent Father in Heaven.

Over the next few years, God began to show me how to hear Him. I realized He'd used the same method of speaking to me at other times in my life, and I looked back and remembered distinct moments when He'd said things to me as a child. He'd spoken directly to my spirit then too.

I had long pictured myself suspended over the Grand Canyon, as though I stood on invisible stepping stones over the abyss. Every time God told me to take a step, my feet landed on something solid, yet when I looked down past my shoes, only the canyon floor gazed back at me a mile down. It aggravated me. I wanted to see the stones before I walked forward, but God said, "Just step anyway." I never fell. Each time I set my foot down, I stood on "nothing" perfectly safe and sound.

The most important lesson came just a few months later. I returned to college for the fall and started trying out different churches. I wandered into a church gym after one Sunday service and played solitary basketball, frustrated at God's silence, at my own weakness. I spent a good ten minutes in there, shooting baskets badly until I was blubbering. About the time I booted a volleyball across the room, a man walked through with his family.

I turned away from them, hiding my damp, red face and trying to act, you know, like a calm person randomly kicking volleyballs. I expected them to pass by, but instead of behaving like a normal well-adjusted individual and ignoring me, this man walked straight across the room toward me. I don't remember his saying a word to his wife and kids; he just approached me in the middle of the basketball court. Then, without asking, he gently wrapped his arms around me. His arms were rough in his sports jacket, but he was solid and warm.

I didn't feel offended or awkward. I accepted this stranger's hug right there in the gym.

"Jesus loves you," he began to tell me. "Jesus loves you. Jesus loves you. Jesus loves you," over and over again.

"Yeah," I thought. "For the Bible tells me so. I've heard the song."

Then he spoke to the deepest grief in my heart. "God is not going to talk to you the way you want Him to. He's given you the Bible as His Word, and He wants you to read it. When you need to hear from Him, that's where He wants you to go. Jesus loves you. Jesus loves you. Jesus loves you."

I wilted with his arms around me. I could have been having boyfriend problems. I could have been afraid I was pregnant. I could have felt horrible because I'd told my best friend I hated her. Any number of things might have caused a young woman to kick volleyballs in a gym after church, bawling by herself. But, some man I'd never met came up, and he nailed it.

Dr. Stillwell listened to the story with compassion as we climbed out of the Grand Canyon in 2011. When I got to this part of the tale, though, Dr. S. growled, "He was just giving you his shtick."

Well, giving me his shtick or not, that stranger nailed a 10-penny galvanized through my greatest struggle. I wanted desperately to know God's love, and I wanted desperately to hear from Him. The man answered both of those. Before he let go of me, I said, "Thank you, Lord. Thanks for that."

"Besides," I told Dr. Stillwell. "He was right. God does talk to me through the Bible."

He does.

Most of the time, the things I read are very general and apply to every human being on the planet. I'll have a question, and oh, there's a verse that answers it. If you read the Bible enough to know what it says, the verses will pop into your mind.

"Should I tell my boss what a jerk he is next time he flips out on me?"

"And a servant of the Lord must not quarrel but be gentle to all…"
2 Timothy 2:24

Oh, that's right. Okay. I have to work at that whole not-quarrelling thing, but okay.

On rare occasion, something will jump out and speak directly to me, a personal moment between me and my Father, the heart of God speaking directly to me through words on a page.

When I was 12-years-old, I read Psalm 32, and God and I had a face-to-face moment:

I will instruct you and teach you in the way you should go;
I will guide you with My eye. Do not be like the horse or like
the mule, Which have no understanding, Which must be
harnessed with bit and bridle, Else they will not come
near you.
Psalm 32:8-9

I read that, and I nodded, "Okay, Lord."

At the time, I didn't know what it meant to be guided by the eye of God. I had this picture of a force field blazing from His eyeball, moving me wherever He wanted by His intense magnetic gaze. As an adult, I had to tell my precious 12-year-old self, "No, honey. It means to watch God, to be always expectant, waiting for His direction. That way, when He glances at something, it causes you to look in the same direction so you know where to go."

But why? Why the Bible? Why couldn't He just *talk* to me?

My brother Baron says, "You know why God never talks? Cuz every time He speaks, things just pop 'poof!' out of nothing! If I created a planet every time I spoke, I'd be careful about my words too." Baron has a sense of humor.

I've thought about it a long time and finally concluded something very practical. There are a lot of spirits in this world, and there are plenty of spirits claiming to be God talking to people. Any spirit that is really from the God of the Bible should confirm the Bible. Not just a verse taken out of context, but the full intent of the whole thing.[1] If you want to know when a bit of theology is screwy, you should already know what the whole Bible says. That way you don't end up committing suicide to catch the spaceship behind Hale Bopp. You don't drink the Flavor-Aid. You don't go around saying you're a god or Cleopatra or other rubbish.

Don't forget, Moses' brother Aaron threw down his rod and it became a serpent, but then Pharaoh's magicians did the same thing.[2] There are a countless number of gods that people worship, and I'm not going to deny that some actively participate in their servants' lives, because I don't know otherwise. People connect with spirit guides and speak to higher powers in various forms. I don't doubt spirits are out there capable of performing signs and wonders, but I don't know those spirits and have no reason to trust them. I'm not interested in serving anybody less than the Most High God, the God

---

1   In 2 Kings 1:9-15, King Ahaziah sends soldiers to go get the prophet Elijah, and twice Elijah calls down fire from heaven to burn them up. Yet, in the New Testament when John and James suggested calling down fire to destroy certain individuals, Jesus said, "I didn't come to destroy lives but to save people!"
(Luke 9:54-56). We avoid big mistakes by knowing the whole Bible well. I feel nervous even writing about how God speaks to me through certain verses, because I don't want people to get the wrong idea. We have to balance every verse against the entire Word of God.
2   Exodus 7:8-13.

of Abraham, Isaac and Jacob. The God of Isaiah and Daniel. He's the One I want to serve.

Every person gets the ultimate liberty, right or wrong, to decide which god he or she will worship, whether Yahweh or Allah or Rah or woodland tree spirits. We all believe what we choose to believe. People can worship themselves. They can serve the planet. *"Choose for yourselves this day whom you will serve…"*[3] As for me and my house, we serve the God of Israel. And the point is, if I say He's the God I serve, then I should know what the Bible says and not be quick to let other spirits entice me.

Another face-to-face moment came when I was 20, when Jeremiah grabbed me:

> *The LORD has appeared of old to me, saying: "Yes, I have loved you with an everlasting love; Therefore with lovingkindness I have drawn you…"*
>
> Jeremiah 31:3

I didn't want to embrace that verse. I sat for the next 10 minutes trying to randomly find another similar verse, to show it was just an accident I'd turned to those words. In the context, God is speaking to Israel. In that moment, though, He was speaking to me, and I fought it.

Didn't I want to be loved? Of course I did! But, I didn't want false hope! I didn't want to believe anything that wasn't absolutely, completely true and dependable.

Over the years, I've learned to believe God. To trust that He really does love me, and it's a kind, patient sort of love. I always expect Him to be cross or disappointed, but when I get a whiff of His presence, I feel His patient kindness toward me.

Another thing I've concluded is that God simply wants me to trust Him. He wants me to believe He is who He says He is, regardless of how things look. The children of Israel saw the Red Sea split in front of them, but they still didn't trust God to protect them in the

---

3   Joshua 24:15.

desert.[4] They saw water pour from a rock, but they didn't trust God to provide for them.[5] And Eve. Eve talked with God every day, but she still believed Satan's lies.[6]

God wanted me to learn to believe that He loved me just because He *said* He did. And I've found that when I trust Him, things consistently work out much better than I ever hoped. That's the evidence of my life; it's what I've found to be true over and over and over again.

Most of the time, years go by and I hear no special words from God. Years. I've learned to trust Him anyway. I trust that He hears me when I pray. I trust Him to move me where He wants me to go. I trust Him to be who He says He is. I think that's hugely important to Him - just to trust in His love for me.

I've spent years studying the Scriptures, reading them through, learning about their history and the meaning of the verses in the context of the time. I read every book as honestly as I know how (even those genealogies in 1 Chronicles), and I've worked through a lot of questions. I'm not afraid of the challenges or anger or confusion people have the Bible. I've often wrestled with the same questions.

The reason I bring all this up is that God started using the Bible to direct me regarding Dr. Stillwell. Not many verses and not very often, but they got my attention when it mattered.

---

4 Exodus 14.
5 Exodus 17.
6 Genesis 3.

Chapter 17
Go Not Forth Hastily

"If you drink enough Coca Cola, you can get a lot of carbonic acid, destroy your kidneys, and get a little alcohol as well." - Dr. Stillwell

One particular morning that spring 2010 semester, I snuggled up in my fluffy recliner at home, reading Proverbs. As I read along in Randy's King James Bible, a particular proverb struck me on the nose:

Go not forth hastily to strive, lest thou know not what to do in the end thereof, when thy neighbour hath put thee to shame.

Proverbs 25:8

I squinted. "I haven't been arguing with him." I looked up at the ceiling. "I haven't been arguing with him!" I hadn't even worried about striving.

As I said, I'd gone to Dr. Stillwell's geology class that spring determined to learn what he knew. I wanted to hear what the general scientific community said about the history of the earth and therefore did not go into Dr. Stillwell's class to argue with him about anything. Geology was his field. He knew it. I didn't. Super simple.

And nothing Dr. Stillwell said really bothered me anyway. As a matter of fact, he tried to show sensitivity to the faiths of people in the classroom. He was a little condescending and clearly wanted to humor certain people, but I appreciated the effort.

"They were trying to explain the world based on the information they had," he referred to the writers of the Bible one day. "And the Bible is useful for people studying archeology." He went on to argue that Noah's Flood may have been an actual event localized in the Mediterranean. I didn't necessarily agree with him, but I didn't see any need to battle him on it.

"If you want to believe in the six days of Creation, you have to stretch those days to millions of years," he suggested another day.

I didn't agree with that at all. The specific Hebrew syntax of those verses, "And it was evening, and it was morning," told me, "This is a day." What a "day" meant in a world with no sun is another issue, but if you want the world to be billions of years old and fit the Bible, I think you have to try something else. Still, I didn't feel the need to argue with him about it. The earth was however old the earth was. And I didn't know how old it was, but I figured the puzzle pieces fit together correctly somehow.

Curiously, Dr. Stillwell loved to use biblical idioms. He'd say things like, "Okay, we're going to gird up our loins and get through the Devonian today," or, "The meek might inherit the earth, but they don't get things done in Washington." His dimples creased a bit when he said these things, as though he delighted over the inside joke with himself.

One day I said to him, "I can't figure you out."

His eyes sparkled in his pleasure at being an enigma.

I continued, "I can't figure out whether you're a liberal Christian or a Humanist."

"Hmmm…" he puzzled about that for a moment. "Maybe I'm a Humanist."

"Because you make biblical allusions all the time."

"Ah! Most people don't pick up on that," he chuckled.

I found Dr. Stillwell to be a reasonable man. He was honest about the fact that geologists were trying to jigsaw together the planet's history with far fewer pieces than they'd like. He struck me as somebody who wanted to teach us what was true, and he wanted to be frank about what geologists did and did not know.

That's why it was so weird God warned me not to argue with him.

I walked into class that same morning, the morning of my Proverbs 25:8 admonition, and sat in my squeaky metal chair next to Dr. Stillwell's computer station. I set one foot up on the chair in front of me and leaned back, arms folded, all comfy and ready to watch Power Point slides and scribble down notes and snicker at the doc's one liners.

> Dr. S.: If the earth starts doing its thing, we won't be here much longer, but it will be spectacular!

He cracked me up.

Wouldn't you know, for the very first time that semester, the good doctor said something that provoked me? We'd started covering the history of Earth from the Hadean Eon to the Archean Eon and onward. Dr. Stillwell had taught us about faults and anticlines and synclines. Then, on *this* particularly day, the professor brought up the Middle Ages and blamed the medieval Church for the rampant ignorance of the times. His implication was that Christianity diminished the general intelligence of the population, and he said it with a bit of vinegar in his voice.

As soon as he made his not-so-subtle jab, a list of brilliant Christian men scrolled through my mind. Robert Boyle, Lord Kelvin, mathematician Bernhard Riemann were all believers. Even Nicolas Steno, the founder of modern stratigraphy and geology, whose principles we'd discussed the very first days of class, even Nicolas Steno was a devout man of faith. These weren't men who just went with the flow, believing everything they were told. They were men with innovative minds, men who cared about what was true.

Jesus didn't pander to anybody. He saved lives. He showed mercy and offered forgiveness and healing, and He didn't compromise whenever some religious leaders of His day threatened Him. His interchanges with the Pharisees are fantastic. Have you read them? Jesus was the man.

Dr. Stillwell's attitude had a nasty bite that morning, and there was nothing balanced about it. It aggravated me, and for the first time that semester, I wanted to get into it with him. If it wasn't for Proverbs 25:8, I'd have gone up to my geology professor after class and made some noise. I'm good at making noise.

Instead, I agreed, "Okay, Lord. I won't argue with him." And I let it go.

What if I'd debated the geologist? I don't know. Dr. Stillwell might have blasted me with the evils of Church History. I might have argued that the institutionalized Church of the Middle Ages didn't represent Jesus Christ very well. I don't know what mess I could have gotten into; I didn't go forth hastily to strive with him.

I get the idea, though, that there was more at stake than a disagreement about medieval religion. Looking back, I think it was just a bad day. I know Dr. S. well enough now to recognize he'd probably had a thorn poking at him. When somebody has a thorn poking at him, it does no good to add to his aggravation.

I don't know, but I've learned that God doesn't direct me if it's not important.

Chapter 18

The Lost Father

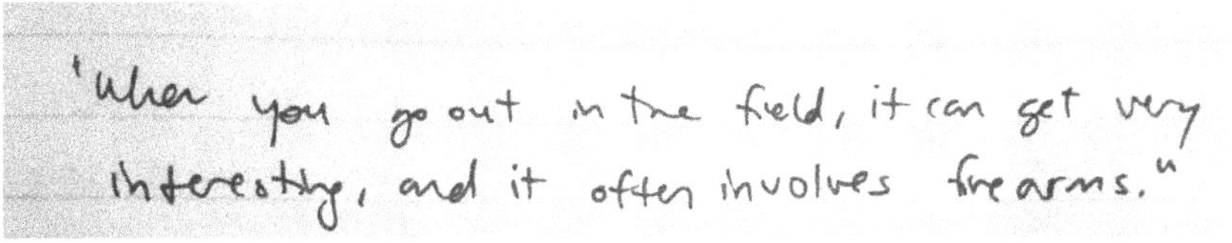

> "When you go out in the field, it can get very interesting, and it often involves firearms."

Figure 7: From a Dr. Stillwell lecture, March 2010.

Dr. Stillwell was a good teacher. His fun personality made it easier to understand and remember the material. He also told stories. It was one of the best things about his class, his occasional tale about driving his ancient VW bus over the Tetons while his brakes failed, or earning peach pie from a rancher who initially greeted him at the door with a shotgun. On rare occasions, the old doctor even told stories on his wife and children.

"My daughter grows out her hair until it's so long and thick and heavy that it gives her headaches. I told her, 'If it gives you headaches, then *cut* it.'"

Another day he boasted, "My daughters and I like to sit and watch Jeopardy together. I always answer questions before they do, and they just hate it." He chuckled at himself.

I gleaned through his stories that his kids were about my age - maybe a few years older. It was easy to see Dr. Stillwell as a father. He treated all of us as though we were children.

He asked me one day, "Did you eat breakfast this morning?"

"Yes, sir."

"What did you have?"

"A peach."

He grimaced. "That's not enough."

It cracked me up. I loved it.

He took our class out to examine road cuts and gather samples near Cumberland, Maryland one Tuesday lab day. "Everybody to the van! We're going to Cucumberland!" We parked beside the freeway next to an exposed section of mountain, and Dr. Stillwell split us into groups and sent us hunting through rocks filled with crinoids and coral and bivalves.

I got frustrated with picking through fallen debris, so I climbed higher to gather samples from the locations they'd been lodged for eons, to pull my rock chunks straight from the mountain. Unstable stones littered the slope, and I took care with each step as I worked my way up the hillside. Before long, I heard a bellow from my professor.

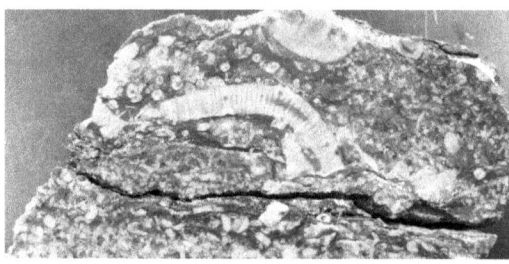

Figure 8: A bit of crinoid stem I found in a rock I sliced open from that day. Dr. Stillwell suggested it was a "lucky cut," but I was pleased.

"WHEN I WAS A SMALL CHILD, my mother used to say, 'If you fall and break both your legs, don't come running to me!'"

I relented and climbed down, but I laughed quietly with my fellow students. "It's funny, don't you see? Because if you fall and break both your legs, you can't run anywhere. Hahaha!!"

Dr. Stillwell wasn't amused. "My fiery German mother didn't think it was funny."

"Was your mother the disciplinarian in your house?" I asked him.

"No." He raised his eyebrows high. "That would be my father."

Ah. I could tell there was some history there.

Dr. Stillwell's wife worked on campus too. She'd earned a reputation as the "Cat Lady" because of her habit of feeding strays. A few of the cats made it into the Stillwell home, and we became familiar with Georgie-girl and Mr. Bigglesworth and Orson the one-eyed cat as regular characters in Dr. Stillwell's tales.

On the day I started to read his PhD thesis, however, I discovered that Dr. Stillwell's spouse was not his first. He offered thanks to his wife right at the beginning of his dissertation on fenestrate bryozoans, but the name of the woman was not the name of the Cat Lady.

Which meant… Dr. Stillwell's kids probably had a stepfather somewhere.

Hmmm.

I wondered about that. I wondered how their family dynamics worked and whether his children got along with their stepmother on one side and their stepfather on the other. Did Dr. S. and his ex-wife cooperate, or had it been a battle for years?

It wasn't my business, but that had never stopped me from puzzling over things. Stepparent relationships are tricky things.

I had stepchildren of my own, and we loved each other. I never tried to force myself on them, to invade their mothers' space. My darling Amber referred to me as her "stepmommy," which I enjoyed, and when I worked as a substitute teacher at her high school, her friends would holler at me, "Hey Amber's mom!!" I never expected my stepchildren to call me "Mom." I'd just hoped they would regard me as somebody more than, "That woman my biological father married." I treated them the way I wanted to be treated, and we got along wonderfully. I was grateful.

So I wondered about the Stillwell family. Had Dr. Stillwell been a single dad until he'd remarried or were his kids raised by their stepfather? Did his daughters consider themselves daddy's girls or did their mom fill their heads with nasty things about him?

I didn't realize there was more to the story.

"Do your kids have a stepfather?" I asked Dr. S. one day in his office. It was out of the blue and uncalled for, but it fell out of my mouth.

"Yes," he said.

"Do they call him, 'Dad'?" I started, but he interrupted me to answer the real question.

"We have met the enemy, and he is us."

"What! You're a *stepfather*?"

He nodded.

"Do they call you 'Dad'?"

"No, they don't," he said lightly. Then he went on to boast about his brilliant stepson who'd gotten a perfect score on his SATs. Sounded like my big brother Lex.

"Do your children all have the same mother?" (One has to ask questions efficiently when prying into professors' lives.)

He nodded.

Just one set of stepchildren and none of his own? I did quick math in my head based on the date of his dissertation and realized his children were teenagers when he'd entered their lives. He hadn't raised them as little kids; he was a late addition.

Wow. That was so weird. He acted like such a father to all of us.

Maybe that was why. Maybe he'd always wanted to be a dad. Or, maybe he'd lost a child. Either way, he took on his stepchildren and called them his own, and he poured all that extra fathering nature inside him onto his students. He taught us to use compasses and took us camping. He told us funny stories and cooked us chili with plenty of vegetables in it. He yelled at us to get down from hillsides and asked us what we'd eaten for breakfast.

I've never asked him why he didn't have children. I'm sure he'd tell me if I asked, but there are some subjects even I have the decency to leave alone.

Still, I appreciated him. I appreciated his willingness to father us. Some of us needed it.

Chapter 19
Kick You in the Tibia

I give bits and pieces of my life in these stories, and they can be disjointed and confusing. When did I live in Texas? When did I live in Seattle? If I was poor, how did we live on a ranch? All the bits are true and they do fit together, but appearances can be deceiving without enough information.

That's how I see the history of the earth. Geologists take classes on minerology and crystal structure, stratigraphy and petrology and geophysics, and each offers us different puzzle pieces. I wanted to learn how to analyze it all. I was after the story that the strata told us, but the school didn't have a geology major. I could read books, and that was good, but the only formal geology training I could get was what Dr. Stillwell offered me.

Remember, the planet has a true history. Its age is however old the earth is, whether we've got it right or not. The animal skeletons preserved in the sandstone and limestone and shale of Earth's skin got there by specific causes. Something happened. Many somethings. The great fun of geology is figuring out those "somethings." Which layers were ripped away by a mud slide or burst dam - or were formed by sediments that settled in the resulting canyons? Sometimes strata are bent and folded and even overturned. Sometimes they're cooked by heat from below or pressure from above or tectonic activity.

Geologists are investigators. They're cold case detectives that use rocks to write Earth's history.

- How far do certain geological formations stretch?
- What rock types make up those formations?

- Strike and dip?
- Rock size and condition?
- Specific minerals?
- Notable fossils?
- Has the rock been torn or bent or folded?
- Was there a stream bed here?
- Dikes or sills from magma squeezing through cracks?
- Volcanic basalts?
- What was the order of events that made these features?
- Do we find outcrops with similar characteristics in other parts of the country or on the other side of the world?

These are some pieces geologists use to tell their tales.

My old boss Joe David was raised Primitive Baptist, and he faithfully attended church every Sunday. He left his squashed cowboy hat and dusty jeans at home, but he kept the fuzzy lamb's wool beard. It plain didn't matter what his religious background was. It didn't matter whether he believed that life was created over six literal days or several billion years, Joe still found Pliocene creatures in the Blanco Formation of West Texas. He knew that if he hunted in that white sand and clay, he might find *Stegomastodon* and little three-toed horses along with hyenas, small camels, giant camels, and sloths. If he hunted in the Dockum Formation - what he called the "Triassic redbeds" - he'd find ancient reptiles called phytosaurs along with huge metoposaurs that look like monster salamanders from some black-and-white Japanese movie.

That's what he found. Those fossils are part of the story of Earth's history. Did phytosaurs turn into tortoises and three-toed horses? No. But, the kinds of animals that lived in that area changed over time for whatever reason. In most locations where fossils are found, it's the same story. Change over time. Kinda like New Jersey.

Before Darwin, William "Strata" Smith came up with an idea

Figure 9: This is the phytosaur skull cast that used to sit on the bookshelves in Joe's living room. It would calmly gaze down on the author as she typed at Joe's computer.

that we should find certain fossils in certain layers, however far those layers extended. It's now called biostratigraphy, and it's one of the methods that geologists use to determine which geologic formation they've found, along with the composition of the rock (limestone? iron-rich sandstone?). Joe knew that when he found beds of red sediments there in West Texas, he was likely to find the skeletal remains of brutally large crocodile- and salamander-like beasts.

"Well, down in those Triassic redbeds, I found a giant waterdog completely articulated," Joe might say. Joe's mix of colloquial Texan and paleontology always entertained me. "I'm fixin' to mold this spinous process."

Of course, camels still exist, they've just changed size and moved elsewhere. Today's horses don't express all three toes like their smaller cousins once did. Salamanders wander all over Texas, but they're a bit smaller than the mastiff-sized ones Joe found in those ancient sediments. The wide-horned *Bison latifrons* doesn't graze on any Texas grassland today, but modern bison and cattle do. Phytosaurs don't turn into three-toed horses, but populations do change.

"Evolution" isn't an evil word. Its fundamental meaning is "change over time." Languages evolve. Music evolves. Populations evolve. The word "evolution" has several connotations in our world today, so before you get into an argument about evolution, make sure to define exactly what you mean by the word.

Figure 10: Samuel with Joe David's cast of a giant Bison latifrons *skull.*

For the most part, natural selection and survival of the fittest are easy to observe. Black panthers might flourish better in one region while spotted leopards survive better in another. Hawaiian honeycreepers with big beaks are better suited to one island where they feed on seeds, and thinner-beaked honeycreepers fare better on another where they drink nectar. Each species has survived to produce cute little babies with the same helpful characteristics that allowed their parents to make it to adulthood. Whichever babies endured to sexual maturity were the ones that passed on their genes, resulting in fatter, stronger beaks on one island or slender, longer beaks on another. These basic concepts are seen all the time in the real world. In this sense, "evolution" is absolutely true and readily observable.

Evidence of speciation is seen today.[1] There are fish in Lake Stechlin, Germany that were once the same species but now swim at different depths. Ursula the evil sea queen wasn't required to BOOM separate two fish lovers by making one prefer colder water - or tolerate less oxygen - than the other. There was simply some variation in the fish gene pool, and all the fish in the middle of the genetic mix out-competed each other or weren't as successful. The ones on the genetic edges were different enough to stay out of each other's way. Whatever the case, the two groups no longer breed with each other and are now considered different species. No problem.

Darwin's *Origin of the Species* is filled with his careful observations.

---

1   For example, see Ohlberger, J. et al. (2008) Temperature-Related Physiological Adaptions Promote Ecological Divergence in a Sympatric Species Pair of Temperate Freshwater Fish, Coregonus spp. *Functional Ecology,* 22(3):501-508. Also see Hilton, Z. et al. (2008) Physiology Underpins Habitat Partitioning in a Sympatric Sister-Species Pair of Intertidal Fishes. *Functional Ecology,* 22(6):1108-1117.

He developed his ideas from what he saw taking place in the animal kingdom, anticipating the discoveries later made in genetics.

However, the fight over "evolution" these days is rarely about whether Darwin's ideas correctly explain the differences we see in finches or fishes. That's not the rumble in the jungle. The real puzzle is whether Darwin's observations can be extrapolated to describe the grand evolution of all life from single-celled organisms billions of years ago. That's the big question, and it's important to answer well.

It used to be that Darwinism ruled in geology/paleontology. Not so much these past 40 years. Nope. Darwinism has been amended by Punctuated Equilibrium. In the 1970s, Stephen Jay Gould and Niles Eldredge developed the Theory of Punctuated Equilibrium to address some concerns with Darwin's phyletic gradualism,[2] and their ideas have since received wide acceptance.

Punctuated Equilibria. Punctuated Equilibria.
Pronounce it correctly or I'll kick you in the tibia.

I always appreciated Stephen Jay Gould. He was willing to challenge the whole scientific world when the prevailing models plain didn't match what paleontologists were finding. When I was young and had even more freckles, Joe David and I sat in the living room of Joe's house and took turns reading Gould's essays in *Natural History Magazine* out loud. The cast of an Apache skull stared at us from the china cabinet, along with mastodon teeth and aetosaur horns, while we enjoyed the Harvard professor's pleasant, easily-digested writing style.

Long before my time, Eldredge and Gould pointed out that if evolution occurred according to Darwin's theory of gradual change, we should see a step-by-step series of creatures in the fossil record. We should be able to lay out the fossils in order as one species slowly morphed into the next in a drooling rainbow of critters over millions of years. That's not what paleontologists actually found, Eldredge and Gould reminded everybody; we were missing the multitude of

---

2   Eldredge, N., and Gould, S.J. (1972). Punctuated Equilibria: An Alternative to Phyletic Gradualism. In Schopf, T.J.M. (ed.), *Models in Paleobiology* (p. 90.). San Francisco: Freeman, Cooper & Co. Also see Gould, S.J. (1989). Punctuated Equilibrium in Fact and Theory. *Journal of Social and Biological Structures*, 12:117-136.

biological links that Darwinism predicted.

Don't be confused. We have a large number of equines in the world: zebras, asses, onagers and draft horses are all members of the genus *Equus*, and mitochondrial DNA evidence shows us how closely ancient species of the family Equidae are related to domesticated horses.[3] There are some chromosomal differences between members of *Equus*, but horses and donkeys and zebras can still interbreed.[4] They can all be shown to be related to our favorite little three-toed friend *Miohippus* and a branching bush of other extinct horses.

It's trickier to trace *Miohippus* to ancient species of rhinoceros. The horse and the rhinoceros and tapir are all placed in the order Perissodactyla, ungulates with odd numbers of toes. We have a big bushy tree of horse fossils and we have a big bushy tree of rhinoceros fossils, but even experts on perissodactyls acknowledge that these families appeared suddenly in the Eocene, and there is little evidence on how they are related to each other.[5] In other words, we don't have fossils that show us how the horse and rhinoceros and tapir evolved from a common odd-toed ancestor. That doesn't mean that we can't do DNA analysis or compare similar and dissimilar biological characteristics between taxa, but it would be nice if paleontologists could provide the actual bones.

This is a common problem in the story of evolution, and the normal answer to the problem is, "The fossil record is incomplete." We don't have every once-existing skeleton at our disposal. Gould and Eldredge knew the normal response, but they groaned that the

---

3   For example: Jansen, T., et al. (2002). Mitochondrial DNA and the Origins of the Domestic Horse. *PNAS*, 99(16):10905-10910.

4   The chromosome count varies among species in *Equus* (horses-64, donkeys-62, and Przewalski's horses in Mongolia-66) but the species can all interbreed. Zebras have even more diverse chromosome numbers (Grevy's zebra-46, the plains zebra including Burchelli's zebra -44, and the mountain zebra -32). Yet, these species can produce viable offspring with horses - the "zorse." Studies show that the centromeres in *Equus* chromosomes have had a tendency to move, allowing certain chromosomes to divide or fuse while the gene orthologs remain highly similar and functional. Cf. Carbone, L., et al. (2006). Evolutionary Movement of Centromeres in Horse, Donkey, and Zebra. *Genomics*, 87(6): 777-782. Doi.org/10.1016/j.ygeno.2005.11.012. // Piras, F., et al. (2022). Molecular Dynamics and Evolution of Centromeres in the Genus Equus. *International Journal of Molecular Sciences*, 23(08): 4183. Doi.org/10.3390/ijms23084183.

5   The American Museum of Natural History *tries*. Its website page on perissodactyls offers uncertainty with courage, declaring, "While we don't know exactly what type of mammal perissodactyls are descended from, we do know that their ancestor was probably very similar to, and possibly even a close relative of, a group of extinct mammals called phenacodontids.https://research.amnh.org/paleontology/perissodactyl/evolution/intro. Last accessed July 16, 2024.

argument still left paleontologists with little skeletal evidence for Darwin's theories, and Darwin's phyletic gradualism threatened to be unfalsifiable.

To be "falsifiable" just means that an idea has to be testable, so that it *can* be shown false if it's wrong. In fact, that's the goal - to prove it false. If test after test after test shows it isn't false, then we can have some confidence in the idea. Once we share it with the rest of the world, other scientists can test it too. If a theory doesn't offer predictions that can be tested, even indirectly, it's considered "unfalsifiable" and therefore not scientific.

Darwinism predicted a vast array of intermediate creatures as evolution took place through the ages. If paleontologists never found the skeletons of all those intermediate forms but refused to let go of Darwin, then Darwinism was no longer falsifiable from a fossil point of view and therefore not scientific.

Darwin's observations along with discoveries in genetics explained the variations of horses in the Equidae bush but not the assumed evolution of the horse from an ancient perissodactyl great great grandma. If horses and rhinoceroses came from a common ancestor, bone diggers should find the evidence as they unearthed skeletons over time. However, decades and centuries passed, and those intermediates never showed up in the fossil record. And obviously it wasn't just perissodactyls. Darwinism predicted that paleontologists would find the gradual changes that led from ancient ancestors to modern forms of creatures all around the world, but they didn't appear.

This lack of fossil evidence shoved a stick in the spokes of microbes-to-man evolutionary theory. A young-earth creationist could have correctly predicted 100 years ago, "If God made the world, then we expect the fossil record will show 'kinds' of animals, according to Genesis 1:24-25, but not the evolution of one 'kind' into another." The evidence supported the creationist position much better than Darwin's in this regard. Darwinists needed to find those intermediate forms or admit Darwin's theory falsified.

Gould got quite forceful about the problem by 1977, writing:

Phyletic gradualism was an a priori assertion from the start - it was never "seen" in the rocks; it expressed the cultural and political biases of 19th century liberalism. Huxley advised Darwin to eschew it as an "unnecessary difficulty." We think that it has now become an empirical fallacy.[6]

We find giant camels and little horses with two extra toes above their fetlocks, we find giant beavers and sloths and enormous gophers with horns. We unearth the mammoths, but we don't find all the required evolutionary steps we'd expect to find if mammoths evolved little by little from a common ancestor with manatees and hyraxes. Not in West Texas anyway.

Gould and Eldredge recognized that additional ideas were necessary. Their Punctuated Equilibrium theory offered an answer to the fossil record problem by suggesting that evolution did not occur gradually, but rather in spurts. Creatures could reproduce generation after generation for millennia, passing on their genes just as they received them from their parents. The genetic code would get a twist in a small group, and evolution would occur suddenly (geologically speaking) in some isolated part of the population. Another long period of conservation would follow, eventually punctuated by another burst of evolutionary development. Those best suited for their environments lived on.

Gould and Eldredge warned that because fossils are only preserved under special circumstances and major evolutionary changes take place in small subunits of the population, we shouldn't expect to find the drooling rainbow of missing links. In other words, Darwinists shouldn't have predicted a multitude of intermediates. We should only find cross sections of the ancient world in our layers.

The fossil record does not provide the host of evolutionary stair-steps we'd expect, but we recognize that time and erosion and circumstance have cheated us from a thorough record of every creature the planet has ever held. The geological camera would rarely have captured evolutionary breakthroughs in small portions of a

---

6   Gould, S.J. and Eldredge, N., (1977). Punctuated Equilibria: The Tempo and Mode of Evolution Reconsidered. *Paleobiology*, 3(2):115.

population. According to Gould and Eldredge, we should expect to see snapshots of long-lasting species.

I liked Gould because he didn't try to force squares into round holes. He had the decency to say, "This doesn't work. Let's look at things a different way." I liked that about him.

And there *is* the whale sequence! For now. Whale evolution is considered supported by fossils from the four-legged land animal *Pakicetus* with a bony wall around its middle ear to *Ambulocetus* with its powerful tail and shorter legs to *Kutchicetus* that liked salt water to *Rodhocetus* that had both land and sea mammal characteristics to *Dorudon* with tail flukes and its nasal opening farther back. It's not a drooling rainbow, and there are anatomical features of whales that aren't explained, but the paleontological community seems content and even pleased with these skeletal samples. So there's that as well.

I look at this whole thing, and this is what I see.

Ultimately, we *do* tend to find certain animals in certain formations. Joe always found his phytosaurs and metoposaurs in the Triassic redbeds of the Dockum Formation. Cretaceous layers in Wyoming and Montana hold T-Rex and *Triceratops* and duck-billed dinosaur remains. This tends to be true.

However, we don't know the whole story of how the swampy Triassic red sandstones gave way to the white, calcium carbonate-rich sands and clays of the Blanco Formation where the camels and little horses played.

And there are strange things, things that don't fit the standard geological model.

I read an informal report that Joe wrote in 1983 in which he describes the tusk of a *Stegomastodon* from the Blanco Formation that he found curled into the red rock of the Dockum Formation below. There's supposed to be 200 million years between those two layers. The Jurassic, Cretaceous, Paleogene - et cetera - layers are completely missing from between the Triassic red mud and the Pliocene white sand in that area, and it was unexpected to find a *Stegomastodon* tusk invading into the hard Triassic rock.

When I asked him, Joe told me he thought a small earthquake had been involved in allowing the tusk to bury itself, and maybe that's all there was to it. On the other hand, maybe there weren't millions of years between them, and that red mud was still soft when the mastodon got buried. Joe's report states that the Triassic material stuck to the tusk, like mud would do. It was like cement, not crumbly broken rock.

It's a sticky note on the wall of my mind.

Ultimately, we *haven't* found the multitudes of evolutionary missing links that show the step-by-step development of families of creatures from their common ancestors. We find plenty of evidence of the evolution of horses and gophers and sloths within their families, but we don't find stepping stones that connect them all to their original proto-mammal mommies. Gould and Eldredge offered one reasonable explanation for this, but is it the best one?

What if horses plain weren't related to rhinoceroses in any way, ever, and *that's* why we haven't found the links between them? Are those skeletons in the whale evolution lineup legitimate stepping stones, or is the whole thing contrived by people desperate to prove what they already believe? What is the true story of what happened?

I'm being serious, because I don't know the answer. My judgment is suspended. I'm still collecting data.

> and clay which are like cement w. ere left attached to the tusk when we plaster coated it.
> 3. The tip of the tusk is still attached and left encased in the red clay surrounding it.
>
> THE STRATA
> 10. The tusk definately in the red clay about thelve inches deep
> 11. The red clay in some places is completly clearly stratifeed and yet nearby mixed and graded together.

Figure 11: Snippets from Joe David's 1983 description of the Stegomastodon *tusk that had buried itself into the Triassic red rock below the white Pliocene materials of the Blanco Formation.*

CHAPTER 20
CALLING HEADS AND TAILS

Baron: …if I weren't so small and if my nose weren't so big.
Shadow: Wait. Do I look small and skinny to you?
Baron: No. You look fine.
Shadow: Right. And you're bigger than I am.
Baron: Oh.
Shadow: You don't like your nose, you think it's big? Look at my nose. Do I have a big nose?
Baron: No, your nose is fine.
Shadow: Exactly. I'm your identical twin. Every time you think you look bad, just look at me - and remember that you're better looking than I am.

Every year when I say, "It's my twin brothers' birthday today," somebody will inevitably say, "I didn't know you were a twin." I then say, "Well, no. I have *two* twin brothers. If I were the twin, I'd have said, 'It's *my* birthday.'" (I think it's funny every time I say it. I'm the only one who laughs, but that's okay.)

Baron and Shadow both entered the Army when they graduated from high school, and Shadow never left. He's now a Sergeant First Class, responsible for training bright young men.

"I have some catch phrases they never want to hear," Shadow

says. "I don't yell at them. I don't have to. When I tell them to do things a certain way, it's for a very specific set of reasons. But, there's always somebody who doesn't follow directions, and so I have to say,

"'Oh. You did it your own way. Hmmm. Did you think about *this* other aspect of the situation?'

"'Oh. No, Sar'nt.'

"'Yes. And did you think of *that* aspect of the situation over there?'

"'Um… no Sar'nt.'

"'I see. Well, that's unfortunate.'"

Nobody ever wants to hear Sar'nt Truman say, "That's unfortunate." Those two words mean that the next however many hours will be a combination of exhausting and painful. Not to be cruel, but for educational value.

"The smoking has the purpose of teaching the lesson," Shadow says. "As soon as they understand the lesson, I stop smoking them."

Once, Shadow made the offhand comment, "My mother has 300 jumps [out of an airplane]. You guys can do this." Later, he heard muffled sounds of dissension in his ranks. He called the men together and said, "Hey. How many of you actually believed me when I said my mother had 300 jumps?"

Three men raised their hands.

"Okay. You guys hang out here. The rest of you, come with me. It's not looking good for the home team."

Nobody ever wants to hear Sar'nt Truman say, "It's not looking good for the home team."

"Sergeant Truman is totally cool," said one of his former students. "Just make sure you do everything he tells you to do and believe everything he says."

Shadow is serious about people taking him at his word. Don't call him a liar. It's not funny.

In middle school, Shadow told the folks in his theater carpool about jumping off The Rock as a kid. We had a set of small cliffs over a swimming hole down the road from our house, and we used to leap off these things. (Sometimes we still leap off of them, but now we weigh more and the water seems harder.) The tallest rock rises about 30 feet above the water. It's not exceptionally impressive cliff diving, but for a 13-year-old kid, there's a definite thrill involved.

Shadow found a picture of his very first rock jump and showed the others in the car.

They said, "What? You actually did that for real?"

That irritated Shadow. "Of course I did it for real!"

That's the trouble with telling people things they consider incredible. There are liars out there. Lots of them. At the same time, just because something sounds unlikely doesn't mean it's not true. And just because something sounds reasonable doesn't mean it *is* true. Truth isn't subject to our limited perspectives.

I want people to know who I am, to know my strengths and weaknesses. I don't want them surprised when my weaknesses show up, and I want them willing to trust in my strengths. I need people to believe me even when I tell them things that sound unbelievable, so I work to tell the truth as precisely as I can. It does get me in trouble, but one of the worst things you can do to me is suggest I'm a liar. I'd rather you called me a smelly slob with thin lips.

That's one reason Dr. Stillwell and I get along, though. We believe each other's stories. Dr. Stillwell occasionally throws out little gems - like being snowed into a tent for two weeks, eating lentils and reading books before burning them page by page to stay warm. He tells his brief tales casually, because he forgets most people have never been snowed into a tent for two weeks.

I don't doubt him one bit, though, because his stories fit what I know about him. When dealing with people, it's wisest to go with the statistics. Life is not fair. It's not balanced. If a coin lands on Heads 24 times and on Tails 20 times, it might be tempting to pick Tails for the next flip, because you expect things to balance out. No. Go with the statistics. It is more likely that the coin will land on Heads. Choose Heads. If it lands on Heads 40 times in a row, absolutely do not pick Tails for good measure.

The same goes with people. If a man only calls you back once out of the six times he's promised, then don't wait by the phone next time. If your mother is always early to functions, but she's late 10 minutes today, she really might be in a ditch. If an honest and intelligent friend tells you something that sounds bizarre, at least hear him out, and if Sergeant Truman smokes you for calling him a liar, don't call him a liar.

So, I believe Dr. Stillwell's stories. The type of guy who drives over the Tetons with bad brakes would be the same sort of guy to get snowed into a tent for two weeks. He could earn a plate of pie from the farmer who met him at the door with a shotgun. No problem. Dr. Stillwell's stories can be extreme, but they are consistent with who he is.

Same with my mother. The woman who shoots coyotes as a little girl is the same type of woman who might jump out of airplanes. She might move her kids into a cabin with no electricity and running water, and absolutely would raise a pack of self-confident children who would jump off cliffs and dig up dinosaurs and wrap up terrorists and run major corporations and argue with Dr. Zenith.

I showed this old trimmed-up Polaroid to Dr. Stillwell one day and said, "Isn't this a great picture!"

He said, "Yes. And it's also very high."

Figure 12: A young Shadow Truman making his first leap off The Rock.

Chapter 21

Mr. Davies and the Fire

"There are more things in heaven and earth, Horatio, than are dreamt of in your philosophy." -*Hamlet* via Dr. Stillwell

I really appreciated Dr. Stillwell's response to the first incredible thing I told him.

The poor man had been out camping with his earth science class that weekend, the weekend of April 10-11, 2010, and he'd tweaked his tricky back. He hobbled about school with a cane on Monday like an old man. I felt pained for him.

He limped into class Tuesday with his cane and tossed exams at the ends of each black lab table. We took them and handed the piles down to each other. Then, he settled in his computer chair, and we settled into our tests.

I eyed our professor with compassion; his hobbling had looked stiff and unpleasant. "How are you doing?" I asked him.

He didn't complain; that's not Dr. Stillwell's style. He said he had an old injury that tightened the muscles in his back, and this was the result if he wasn't careful.

I don't know how I dared to say it, but it came out of my mouth anyway. "I wish…I wish I could pray for you and have you healed."

He surprised me by nodding in agreement.

"But people don't get healed when I pray."

That got a smile out of him.

I offered lightly, "My husband prayed for people, though, and they were healed."

I was supposed to be taking a lab exam - *an exam*, good grief - so I returned to scribbling out answers on my paper.

It didn't take long before the conversation continued.

I turned back around to him. "Do you want to hear about a miracle?"

Dr. Stillwell nodded, so I told him a story.

"When we were growing up, my mom always told us about Mr. Davies, this old black man who lived on a hill a few miles from her house. They would ride their horses up there and talk with him, and he'd tell them about Jesus. He had a little shack of a house that he'd built from bits of this and that, and he had a few chickens and goats and lived up there by himself."

My mother repeated this story to us kids for years. I told Dr. Stillwell, "But this past Christmas when she told the same story to her brother Doug, he said, 'You dummy. I was there!'"

I never knew Uncle Doug had been a witness to the Mr. Davies miracle.

"I called my Uncle Doug a couple of days ago and asked him about it. He's six years older than Mom, and he was able to give me details Mom was probably too young to know. She was only nine-years-old when it happened, but Doug was 15."

Uncle Doug answered my questions with matter-of-fact calmness. He described the same things Mom had always told about, but better.

"They grew up in the desert north of Los Angeles, and it rained a lot this particular year, so the grass grew super high. Then, the Santa Ana winds came like they do in the late fall, hot and dry, filling the air with static. Doug and Grandpa saw lightning hit up in Iron Canyon, and so locally they called it the Iron Canyon fire. With all that tall grass, the fire raged through the area. You know how those wildfires go. Mom had friends who lost their houses, even with all the fire fighters trying to protect them."

Dr. Stillwell nodded. He knew how those wildfires went.

They lived in what is now called Santa Clarita. I found data on higher-than-normal rainfall in 1958, and the *Los Angeles Times* December 2, 1958 front page roared "Malibu Blaze Running Wild."

The news report tells of a number of fires starting in various canyons and joining together. The news might have been focused on Malibu, but about that same time major fires roared through the little desert communities 45 miles to the northeast. Uncle Doug said the fire that he witnessed burned clear out to Palmdale.

I kept my voice low, because the other students were actually working. "When the fires were gone, their parents told them not to go out on their horses because of the hot spots; they would burn the horses' feet. But, Mom was really worried about Mr. Davies, so she and her friend Toni Sullivan rode out to find him anyway. They hoped that Mr. Davies had gotten out in time."

The charred landscape stretched black and barren in all directions. Mom and Toni picked their way up to the top of Mr. Davies' hill expecting to find the hilltop burned black, the shack gone. But when they reached the top, their horses suddenly waded through grass chest high. Instead of smoldering ruins, there stood Mr. Davies' little shack and his chickens and goats with a circle of grass around them about 50 yards in diameter.

"They said, 'Mr. Davies! What happened, what did you do? How did you stop the fire?'

"He said, 'I saw the fire coming and got on my knees and asked God to spare me, and the fire just split and went around my house and then joined up again over there on the other side.'

"Mom and Doug both said the same thing. They both said they sat there on their horses with the grass up to their horses' chests and stared at the blackened landscape as far as they could see with that little oasis of grass around Mr. Davies' shack."

Mom always reaches out her arm and points when she tells the story. She says, "It was just from here to there. When you got out on those mountains, you could see ranges. You could see the valleys and other hilltops, and it was black everywhere as far as we could see. The only bit of life was the dry grass around his little old shack he'd put together from pieces of this and that."

"He had a pump to get his water," Doug said, "He'd always say to us, 'Now, don't try to use that pump. Its handle is broken, and I have to pump it a special way.'"

As I told this story, Dr. Stillwell didn't laugh or roll his eyes. He didn't turn away in disgust. He said an extremely reasonable thing. He asked, "Is there an alternate explanation?"

I shrugged. "I don't know. Those winds. Doug said that he and his dad were driving on the road beside the fire, and it was moving as fast as their truck."

"Those wildfires do move fast," Dr. Stillwell nodded.

"And he was on a hilltop, where the wind blows hardest. What could that old man have done to protect himself when even the firefighters hadn't been able to save people's houses down below?"

Dr. Stillwell nodded and shrugged. "Maybe it really was a miracle."

I nodded too.

Then Dr. Stillwell offered another side to the issue.

"But what about Grandma?" He invented a hypothetical old woman for the purpose of argument. "When Grandma prays to be healed and she isn't healed, what do you say to her? Now, she feels like God doesn't love her. What about Grandma?"

It was a serious moment as we both recognized the reality of the pain we all experience as human beings. It's always there, all around us.

I looked at him. "I prayed and prayed for my husband to be healed, and he wasn't, and God let him die. He *let* him die, leaving me a widow and my children without a father." I gazed earnestly at him. There was no softening that reality. "But, you know what? It's…it's okay."

I didn't know how to explain it to him.

"It's okay," I finished.

I still missed my husband and my heart still hurt, but I was valued and protected and safe. God hadn't dropped me or my little kids. I felt His presence stronger than ever since Randy's death, and deep inside I felt solid. I knew it was going to be all right.

God saved Mr. Davies' little shack, but he didn't save the homes of those people down in the valley. People had been healed when Randy prayed, but Randy himself died. What do you say to Grandma?

I was glad Dr. Stillwell listened to me and didn't roll his eyes. I truly appreciated him for that.

What about Grandma? What do we say to her? What do we say to the suffering people who think God doesn't care?

Chapter 22
Science and the Spiritual

That very first day I met Dr. Stillwell and watched him hunch over the papers on his black lab table, I thought, "He doesn't know. He doesn't know we're going to be good friends." I sensed it there that first day. He sent me off with Dr. Zenith, because he didn't know. Neither of us understood, though, how much trouble I would cause him because I believed the spiritual world was real - really real - while he did not.

Many years ago, I sat on the Greyhound bus with a handsome metaphysical naturalist. We engaged in a friendly conversation about the nature of Reality, as young people tend to do. I remember his looking up around the bus and asking, "Then where is it?" He grabbed at the air with his fingers. "Where is the spiritual world? I can't see it. I can't touch it. Where is it?"

It was a good question. I puzzled about it for years after, and I've decided theoretical physics offers the best answer. I don't think what we call the "spiritual" is some magical fairy dust place. Physicist Edward Witten's M Theory suggests there are up to 11 dimensions,[1] and I'm convinced the "spiritual" involves additional dimensions past the ones we see with our eyes (wound up as superstrings or otherwise). That's my short answer to the handsome young man's question. We can't access the spiritual world through the four dimensions of space-time that we directly experience, but that doesn't mean it's not there.

Dr. Stillwell recognizes that we're spiritual beings. He says he feeds his spirit in the woods, hiking under the trees and enjoying nature.

---

1   Edward Witten offered M Theory at a USC conference in 1995 to unify various versions of string theory. The "M" is most often said to stand for "mystery."

That's what he says. I suspect his version of "spiritual" has more to do with brain chemistry than with a fifth or sixth dimensional reality, but he does sense an inner part of himself.

What are good scientists supposed to do about spiritual things? On one hand, there are people whose imaginations get away from them, who claim they saw an angel because some kind stranger gave them five bucks for gas. There are those who do anything for attention. There are those who hear voices. We've learned that bacterial infections and mental illnesses cause certain problems once attributed to demonic possession.

On the other hand, if a little boy starts cursing in ancient Sumerian and levitating… the devil might be in the details.

In all honesty, how should a good scientist deal with the spiritual?

What we call the "spiritual" or the "spirit realm" is the domain of religion, while science is firmly stuck in the physical realm. It's not a matter of opinion; scientific method can only be used on the observable physical world. It has only natural explanations in its toolbox. We can't take our spirits and put them in culture tubes and run chemical tests on them. We can't capture other people's sixth senses or set up conditions where we see the same miracles repeated time and again according to some push-button spiritual law. Not easily, anyway. It's almost as bad as trying to capture dark matter. There are places science has a hard time going.

But here's where many scientists make a terrible logical mistake. They erroneously leap from the self-evident statement, "Science has to depend on observable physical evidence" to the unprovable belief that, "The observable physical world is all there *is*." There's a magnificent distinction between those two statements, and it's important that we all recognize it.

In his 1991 book *The Holographic Universe*, author Michael Talbot wrestles to make sense of alleged spiritual manifestations through the discoveries of 20th century science. He considers a bizarre range of documented "miracles" from the amazing power of the placebo effect to the activities of poltergeists. He describes kahunas in Hawaii who walk on burning rocks without harm, and convulsionaries in 18th century France who showed no signs of bruises or injuries after having been severely beaten and tortured. He relates an incident in which a man under hypnosis was able to see through his daughter as though she were not there.

Talbot brings up a variety of mysteries that may or may not be elaborate hoaxes or legends or old wives tales, and he treats them as real events. His explanation is that we live in a holographic universe and our minds are holographic in nature, able to control our own bodies and the world around us. It's an interesting idea.

Then, on page 149 of *The Holographic Universe,* Talbot comes out of the closet. He explains that he is tackling this weird stuff because from early childhood on, he experienced poltergeist activity himself.

> My mother tells me that even when I was a toddler pots and pans had already begun to jump inexplicably from the middle of the kitchen table to the floor.[2]

I laughed out loud when I read that. All of Talbot's talk about Fourier wave forms and quantum potential, and his book came down to a guy trying his darndest to explain the paranormal in scientific terms because he'd watched gravel shower from his own ceiling. He had dreams about events before they happened and knew information about people he had no right knowing. His personal poltergeist materialized things; it once even threw spaghetti on him. He spent years scouring theoretical physics for an answer to it all. He says:

> Since our poltergeist left my family's home and followed me when I went away to college, and since its activity very definitely seemed connected to my moods - its antics becoming more malicious when I was angry or my spirits were low, and more impish and whimsical when my mood was brighter - I have always accepted the idea that poltergeists are manifestations of the unconscious psychokinetic ability of the person around whom they are most active.[3]

---

2 Talbot, Michael (2011). *The Holographic Universe* (p. 149). New York: Harper Perennial.
3 Ibid.

Talbot decided that the incidents were caused by him, by his own mind. He developed his ideas about the holographic nature of the universe in his effort to find a logical, concrete explanation for the bizarre things he experienced. He had physical evidence - gravel and spaghetti and rocking pots - and he also insisted on a solely naturalistic explanation for all of it.

Sadly, Talbot died at the young age of 38. According to the Forward of the 2011 reprint of his book, he passed away from chronic lymphocytic leukemia in May of 1992, and he never knew how well *The Holographic Universe* sold. The very fact of his early death, though, raises a critique of his arguments. He wasn't the normal, thick-skulled mortal; if it were up to him and his sensitive mental powers, couldn't he have healed himself? If he controlled his body and the world around him with his holographic mind, why did he lose his life to leukemia?

I'm not criticizing him; I appreciate his honest pursuit of understanding, and I wish I could sit at his bedside and talk to him. I'm sad he died.

Talbot's story is sobering to me. I too have known things I have no physical way of knowing, sometimes about people clear across the country. I've been instantly healed, and I've seen other people instantly healed. I've experienced divine direction and warnings. Talbot offered an explanation for his experiences from the magisterium of science, and I offer an explanation from the magisterium of religion. My explanation is that I'm connecting to the Spirit of God who knows all things.

I think Talbot was wrong to dismiss the rest of the spiritual world, the one that interacts with us. Perhaps his poltergeist companion *wasn't* his own psychokinetic ability. I suggest it was an actual entity, something outside himself, and it affected his moods and not the other way around. I believe our minds are far more powerful than we realize, but I am also convinced there are spiritual beings in this world who aren't *us*.

It's hard to use science to understand the spiritual, but it's short-sighted to lightly dismiss the supernatural experiences of people around the world. There's too much that requires an explanation to just toss away these things like spilled salt.[4]

Which of us is on the right track, though? Is Talbot right to search for a purely physical answer to the phenomena he experienced, or am I right to offer a spiritual explanation? The true answer falls into one of those categories, and maybe it falls into both. When we get to these borderlands where the magisteria of science and religion meet, we need as much light as possible to see what's closest to the essential truth.

My goal is to gather as much information as I can. I can't put the spiritual in a test tube, but I can make observations. I can pay attention and watch where the evidence leads.

I don't want to believe a bunch of garbage. I can't depend on garbage. I can only depend on what is true.

There may be confusing gaps in our half-finished picture of Reality because we don't have all the puzzle pieces. But, if we did, the truths understood by science and the truths understood by religion would fit together in perfect harmony. We cannot have religious realities that completely contradict scientific realities. Where there are clear contradictions, somebody missed something.

---

4   I'm having a little fun. Tossing spilled salt over one's left shoulder is an old superstitious act, symbolic of throwing salt in the Devil's eyes…

Chapter 23
Matthew

One day, Matthew Caerphilly arrived at school with his jaw wired shut. He jaunted up the sidewalk and grinned at me, rubber bands visibly holding his teeth together.

"Who punched you?" I asked.

He grunted through clenched teeth, "You heard?"

I jolted. "No, I was just joking! Somebody really punched you?"

At this point in the story, Matthew was a goofy 19-year-old who ate lunch with me and the other girls. I found him annoying at first, this gangly kid with hair that stuck straight up all over his head like a blonde Chia Pet. When he cut it short, it was okay, but he looked like a dandelion gone to seed if he let it grow out. He had bulgy blue eyes and a ridiculously loud laugh, and he answered questions all the time in Dr. Dan's chemistry class. I learned to put up with his youthful enthusiasm, though; Chia Pet Boy was a good sort of fellow.

In fact, it turned out he'd gotten belted in the face because he'd tried to rescue a damsel in distress. He saw a woman running from an enraged man and stepped forward to see what was going on. He shouted, "Hey!"

In response, the brutal fellow marched over, decked Matthew and broke his jaw. "Somebody obviously needed self-defense lessons," Matthew later teased himself.

Chia Pet Boy ended up having far more value in my life than I ever suspected. He became my side kick and best pal, my co-conspirator in pranks and puzzles. His parents eventually adopted my fatherless family into their lives, and Matthew introduced me as his sister. Our friendship lasted long past our school years and will likely continue til we're old and decrepit.

That's all for now. Introductions.

"World, meet Matthew. (Matthew, offer greetings to the world.)"

"Greetings, world."

In the fall of 2010, it mattered most that Matthew served a vital role as my physics partner, my armor bearer in the mortal struggle I waged against the self-important, ever-melodramatic astrophysicist Dr. Zenith.

Chapter 24
The Astrophysicist

I had no classes with Dr. Zenith that first year. Not one, but this simple fact didn't spare me. Sometimes I had to talk to him, and it always went badly. It seemed the astrophysicist had not forgiven me for rebuffing his advice the previous summer.

He remained professional but generally talked down to me and refused to be helpful. Let's say he and I happened to walk down the hall at the same time. Let's say I asked, "Dr. Zenith, if we have a chili cook-off next year, are you willing to participate?" Dr. Zenith would respond by saying, "I do not deal in hypotheticals."

That's how he was. He'd leave me standing in the hall, blinking.

One day that spring semester, Dr. Zenith bopped into Dr. Stillwell's lab toward the end of our class. While he hovered, I asked him a physics-related question. He gave me a short answer, and after I pressed for more details, he said, "When you are in graduate school and you're taking graduate level classes in astrophysics, then maybe we can have this conversation."

I know my eyes widened. I didn't say anything, but I'm sure my eyes shone, and not with astrophysicist worship.

For the most part, I tried to avoid Dr. Zenith. I drew out the juices of theatric superiority in the physics professor; he clearly did not regard me as a worthy individual.

Advisor time came again, and I had a significant decision to make. I had to take a physics class, and I wanted to take it with Matthew Caerphilly the math genius. Matthew needed Dr. Zenith's calculus-based physics as a requirement for his engineering degree. Of course, I wanted Dr. Zenith's daily attention like a hammer on my thumb.

Except. Except that Dr. Zenith turned out to be one of the most beloved teachers in the school.

"Dr. Zenith is hilarious," people said over and over.

"He does a great job teaching the material."

"His class is the best one I have this semester - maybe *of all time*."

Dr. Zenith? Dr. Dark-Side cookies? I told them, "You guys know there's no milk on the Dark Side, right?"

Groan…I didn't doubt them; I'd already caught a taste of it from a distance. Dr. Zenith stood in Dr. Stillwell's office doorway one day joking loudly with the geology professor. I'd glimpsed them carrying on from down the hall and felt telescoped back to my childhood, watching the big kids have fun together while I was left out. I wanted to be part of the big kids.

Dr. Zenith's class also fit most conveniently into my schedule, and I had a tough choice to make.

"If I take Dr. Zenith, I'll just have to be humble. I'll have to endure his brutal treatment and be patient. That's all."

It was like the time I decided to do a flip off the Road Rock into the swimming hole. I was scared, but then I told myself, "It's only pain." Even if I messed up badly, I wouldn't kill myself; it would just hurt a lot.

I thought about that. "Oh. Okay."

This reasoning gave me the courage to do the flip, which ended as a veritable belly flop from 15 feet in the air. It was like hitting concrete - the world exploded orange around me. I floated there in the water, groaning and laughing at myself. Breathe in. Owww. Breathe out. Owww. Take each breath with agonized care. That's what Dr. Zenith's class would be like. Just pain. No biggie.

Dr. Stillwell wasn't as ready to ignore my welfare, and he tried to discourage me from obvious self-harm. He didn't say, "Amy Joy, are you insane? You and Dr. Zenith will attack each other with machetes, and you will lose. Don't do it."

No, Dr. Stillwell wouldn't say that.

He did say, "You know… Dr. *Gurden* is a good teacher. You should take physics with him."

I knew that Dr. Jeff "Flash" Gurden's students often filled the hallway doing strange experiments involving marbles and timers, which I thought looked fun. I had also heard complaints about his class. They all said, "He doesn't know how to teach! He doesn't explain anything in terms we can even understand. He's too new. He doesn't remember we're not in grad school."

I would later develop a warm friendship with Dr. Gurden and his wife. They're near my age. We'd hang out, and I'd call him by his first name. I'd watch his cats when he and his wife wanted to leave town, and they'd watch my kids in a pinch. When I took his Physics II class the Spring 2011 semester, I found him to be a good teacher and learned a lot from him. He's both brilliant and personable, and what's more, he makes the perfect roasted marshmallow. At this point in our lives, though, he was still a new professor and I didn't know him.

Dr. Zenith, on the other hand, Dr. Zenith's students loved him. Yes. They loved him.

It appeared that Dr. Zenith had mastered the secret art of physics education like the great kung fu warriors of old, while Dr. Gurden hadn't broken out of his karate white belt. Wisdom demanded that I benefit from Dr. Zenith's honed skills, even if I had to take a daily drubbing in the process.

So, I went to Dr. Manchester my advisor and signed up for calculus-based physics.

"Okay, let's look at your grades." Dr. Manchester opened my folder and glanced over my mid-semester report. I peeked and saw a stream of As and Bs. He slammed it shut again and joked, "Well, that was boring. You'll bring those up, won't you?"

I nodded. I would. I would bring those up.

He glanced down my list of classes for the Fall. "Physics, Cell Biology, Computers in Science, Organic Chemistry…What is this?" He spied a half-hidden scribble at the bottom of my class list. "Physical Geology? You already have a full schedule."

Schhhedule. He's British, so Dr. Manchester says, "Shedjule."

I hunched in embarrassment. "It's fun for me. Because I love geology."

"You know," he said. "I have students who sometimes do Shakespeare with the Rude Mechanicals or they take some light classes, like fencing or archery. It's good to have a bit of a diversion. But, I don't want you pushing yourself too hard. I don't want to find you in an alley one day, drooling out of the side of your mouth and having to say, 'Oh dear, here's my friend Amy Joy sitting in the gutter.'"

I grinned.

"Raise your right hand," Dr. Manchester ordered.

I did as he said.

"Repeat after me, 'I swear.'"

"I swear."

"'That if this class hurts my other grades, I will drop it.'"

"That this class will not hurt my other grades."

Dr. Manchester grimaced, but he gave me my little sticker thingy with the PIN number on it so I could sign up for classes.

I thanked him and hopped up to scoop a handful of licorice bits from his candy bowl.

He watched me. "Don't be shy. Have some candy before you go."

Yes, I had a full schhhhedule; I wanted to pack 19 credits of heavy science classes into the 2010 fall semester. The real issue wasn't geology, though. That next fall I would have to survive the astrophysicist.

This is my warning to you. I'm getting months ahead of myself.

Chapter 25
Baa Baa Black Sheep

Lance (child): I can give you some advice.
Baron (adult): Oh yeah? What's that?
Lance: If you ask God to forgive you for your sins,
    He'll forgive you.
Baron: Okay.
Lance: That's all the assistance I can give you.

The first time I ever entered Al Travis' house, I wondered what my mother was thinking taking us over there. A chunky hole gaped from the wall just inside the front door where somebody had punched through the drywall in a bit of bad humor. Three or five times. The lights only dimly lit the kitchen, and the living room air smelled of cigarette smoke and old bacon grease. Whenever I enter a house rank with cigarette residue, I think of Al's ancient yellow Formica countertops, weeds in the backyard and an empty refrigerator. I think of the holes in Al's wall and the black eye on his adult daughter.

Mom had met Al at Denny's restaurant. The two of them held interesting discussions over coffee at a time in her life when she had few sources of stimulating conversation. As a child, the things I noticed about Al were his bony body and the thick veins in his skinny arms. He had a mop of blonde hair and a face with no fat on it. He was all eyes and cheek bones.

Al had a temper, and everybody knew it. Mom asked him to watch us a few times after school until she got home from work, but I'm sure she didn't have a lot of options. There was no safety or comfort with Al around, and I spent the whole time waiting to see what he would do. My brother Lex had left home by then, so it was just me and the little kids.

Al never went after me. He didn't even seem to get angry at Baron or Shadow, the destructo-twins. No, Al always picked on Heather. Heather, the sick one. Heather, whose whole world looked like the reflection on the inside of a spoon until she finally got big fat glasses, who suffered severe headaches her entire childhood. Heather, who threw up every other day the first three years of her life, who'd knocked out her front baby teeth at six-months-old and struggled to get people to understand her through her slurring, back-teeth S's speech impediment. I could hardly invent a more pitiful child than my little sister Heather.

Heather returned from a birthday party one afternoon shortly after she'd turned seven, and we messed around in the driveway, playing the favorite game where the older sibling ties up the younger sibling and tells her to escape. I tried to hang Heather upside down, but she hollered that the ropes hurt her ankles, so I untied her. I felt bad, and I got a big silver bowl of warm water to let her soak her feet.

Just as I helped her sit down and step in up to her ankles, Al yelled from inside. "Heather!!"

Heather took a few moments to get up, her legs still in the bowl of water.

Al emerged from the house and barked, "Heather! Get in here now!"

I don't know what had set him off. I can't imagine she'd had time to do anything wrong between arriving home from the birthday party and getting tied up by me. Heather stumbled into the house, and I finished cleaning up the rope and the bowl. Then I headed through the door and downstairs into the rec room.

I turned the corner, and there was Al in the big room. He had his belt off and was wailing on Heather, my seven-year-old little sister. My sick little sister.

Fury swallowed me up. I jumped for that strap and hung onto

it. He held on tight above me, and I felt like a little dog snarling on the other end, heaving with all my might while the bunk beds swung past me. We tussled for a moment there in the big room in an angry game of tug-of-war until he finally forced me off.

I expected him to turn on me, but he didn't. I was 10-years-old. I had no real power against him, but he didn't punish me for fighting him. Instead, he left the room. As soon as he disappeared, I said, "C'mon, kids. You all need to go outside." I helped shove my little brothers and sisters out the basement window, then I went upstairs and called Mom at work.

"He's hitting Heather," I sobbed into the phone. "You have to come home!" And she did.

It wasn't the first time Al had mistreated little Heather, and it raised a fury in my young heart. He could have beaten me all he wanted, and I would have gloried in it. It took a lot to hurt me, and I'd have worn the abuse like a gold medal, like Tom Sawyer taking the cane for Becky Thatcher or the Scarlet Pimpernel rescuing his wife during the French Revolution. It horrified me that he'd hurt poor Heather with her headaches and vomiting and awful slurring speech.

The next day I trudged up the hill of our cul-de-sac, still enraged by the whole thing. You know that kind of anger, the kind that aches through your whole body and burns in your stomach and doesn't go away. I wanted to bawl thinking about it. I walked up the hill, storming inside.

That's the moment in my childhood God first spoke to me. I was hiking up the hill - right up the middle of the road - watching the pavement pass under my tennis shoes. I raged to myself, hot with the injustice of it. About halfway up the hill, the Holy Spirit spoke up. Clearly and distinctly, He said to my spirit, "You need to forgive him."

I stopped right there in the middle of the road. I heard Him, and I stopped. Not only did I hear Him, I felt in my deepest soul that He was right.

I groaned and wrestled for a few moments over it. Was I willing to forgive Al?

I thought about it, and I knew I had to. It didn't matter how angry I was. I didn't try to make myself *feel* forgiving. I didn't try any tricks to convince myself Al's behavior wasn't so bad, because of course it was bad. That wasn't the point. I had to do it anyway.

"Okay," I said simply. "I'll forgive him." All I did was agree.

As soon as I said the words, the burden of anger lifted. It physically released its hold, and I felt it go. Even at 10-years-old, I immediately realized that I'd been given a gift. I had been freed of those heavy, caustic emotions, and I finished hiking up our hill a hundred percent lighter in my spirit. The sadness didn't go away, and I certainly didn't trust Al. Still, I'd been freed, and that moment marked itself on me forever.

Most importantly, the incident taught me something about forgiveness. I didn't forgive Al because he had earned the right to be forgiven. That had nothing to do with it; I forgave him because the Spirit of God wanted me to. He didn't want me carrying around that anger.

It also taught me something about how God works. It's interesting to me that I had a choice over whether I would obey and hand over my right to be angry. I had a choice. As soon as I agreed, the anger fell off me like a suffocating blanket. I couldn't have freed myself of that rage on my own; it was taken away. I was *rescued* from it. But, I'd had the option to hold onto it.

Throughout my life, God has always given me the freedom to decide whether or not I'll do things His way. I try to obey Him because He's God, and His way always turns out to be *so* smart after all. But, He gives me an option.

I used to think that God's will was always done. He's God, after all. He's all-powerful. I thought, "Well, of course His will is done." Now… now I don't think so, otherwise, why would we need to pray, "Thy will be done?" God can be all-powerful and He can be good and the world can still be rotten, because He really does let us make our decisions. He doesn't take that from us.

According to Genesis, God created man in His own image. People argue about what that means, and here's my take on it. Among other things, He created us to be autonomous. We're not robots. We're free-thinking creatures with the ability to invent and create, just like God. When somebody says, "If God intended man to walk, He wouldn't have invented roller skates," they're being funny. But, the thing is that God didn't invent roller skates. A human being invented roller skates. Human beings invented apple pie and pizza, guitars and rifles, space shuttles and atomic weapons. God gave us the precious freedom to think for ourselves, to imagine and create and build. It's a giant, risky, ultimately important gift, and I'm grateful to Him for it. He lets us make our choices, even if they are not His, and He doesn't take it away from us, even when we do the worst with the incredible gift He gave us, even when we grieve His heart beyond comprehension. He lets us live with the decisions we make and learn from our mistakes and the mistakes of others. When we listen to Him, though, He heals us and frees us.

> *"Do I have any pleasure at all that the wicked should die?" says the Lord GOD, "and not that he should turn from his ways and live?"*
>
> Ezekiel 18:23

We like it when bad guys get punished. We all feel the justice of it. There's a scene at the beginning of the Jamie Foxx movie *Django Unchained* where several slaves are unexpectedly freed in the woods at night. They're given the opportunity to either save the slave trader that held them captive or blast him away with a shotgun. As they toss off their blankets, we see the scars on their backs, and we know there's no mercy for the jerk trapped under his dead horse. The men trudge calmly over to the slave trader pinned on the ground, and as he begs for mercy, people in the room start to chuckle. Nobody feels sorry for the lowlife slave trader. He made his living on the misery of others, on the debasement of human life. Boom. The shotgun blows him apart, and the audience is pleased. Justice satisfies something deep inside us.

But. There's a point where we all need mercy. We all need it. As

James the brother of Jesus wrote: "*For judgment is without mercy to the one who has shown no mercy. Mercy triumphs over judgment.*"[1] I am willing to show others the mercy that I want for myself.

Pretty much every horrible thing has happened to people in my immediate family. People I love have been beaten up and knocked down stairs and abused in various ways. We've had people rob us and con us. We've had to deal with entire communities turning against us, lying about us in court, false arrest and kidnapping and you name it. Yes, even kidnapping. We've been belted by the worst things humans can do to each other, but God has never abandoned us in any of it.

I learned from my mother's bad decisions. I learned that they screwed up everybody's lives, not just hers. I also learned that when we turned to God and put ourselves in His hands, He took care of things. He didn't let the bad guys win. Over and over again I've seen it: God protects us and saves us and feeds us and heals us. And sometimes, He rescues the bad guys too. And I'm glad.

I didn't trust Al, but I forgave him. I was freed. I later realized that God, God Himself, wanted to forgive Al. He didn't want any of his sins held against him. The violence and anger, the drugs and fornications and who knows what all, God wanted Al released from all of it. It was His desire that Al be free, just as it was His desire that I be freed from my fury against him. It had nothing to do with whether Al deserved mercy, because he didn't. It had to do with who God is and how He does things.

Al had been an atheist when he and Mom met, but to this day she'll say, "I miss my friend Al. Even when he was an atheist, he was one of the few people I could sit with and discuss the biggest questions of life."

Al would talk things through with Mom. They'd pour over the Bible to see what it said. He'd say things he thought would horrify her, but she wasn't ever horrified.

He said, "If God wants me to believe in him, He's going to have

---

1   James 2:13

to reveal Himself to me."

Mom didn't argue with him. "No, you're right."

One day Mom told Al, "You're a sheep."

He said, "What? Why do you say that?"

"Because you ask sheep questions. You say sheep things. You're a black sheep like me, but you're a sheep."

I'm sure Al thought, "H'yeah, right."

Not only did Al eventually surrender his life to the Savior of the world, but all his old drug buddies down at Denny's got saved too. Instead of whatever they used to talk about, they'd sit together over their coffee and talk about the love and goodness of God.

That's amazing, really. All those ornery tough guys.

When Eric Prudehomme heard that Al had become a Christian, he said, "Oh no. Not Al. Anybody but Al." It was Eric, though, who eventually picked Al up for church when Al's lung cancer had progressed too far for him to drive.

"Did we ever decide," Mom asked Al shortly before he died, "did we decide whether we go to Heaven right away or whether we sleep until everybody is resurrected?"

Al shrugged. "I don't know. All I know is that the first person I'm going to see after I die is Jesus." He looked at Mom and said, "Thank you, little girl. Thank you for telling me about Jesus."

When I heard that Al had died, nothing in me felt hostility toward him. I wanted him to go to Heaven. I wanted God to forgive him. When I pass through the same curtain, I expect to see him, and he'll be whole and healthy and perfect. He'll be the beautiful person God always meant him to be, and I'm happy about it.

We human beings are so screwed up. We think that good people go to Heaven and bad people go to Hell. The problem is that there aren't any people that good. Not any. I know that if I tried to stand before God on Judgment Day and tell Him how good I've been and how I deserve eternal life… I wouldn't. I wouldn't even try. God and I both know me too well for that. We have to be completely perfect to get into Heaven, and I don't qualify. What's it matter to the orchard owner if an apple is 15% bad or 70% bad? God can't let us in if we're even a little tainted, because we'd corrupt everything and then Heaven wouldn't be Heaven. It would be Earth forever, and that would be awful. That's why God made a way for us to be made brand new.

Our sins deserve to be punished. Justice demands it. Jesus didn't sweat blood there in the Garden of Gethsemane, agonizing, because He had some silly, inconsequential job to do. If there was any other way, He wanted God the Father to find it. Yet, in the end He said, "Thy will be done." He had a choice too.

I'm glad that God wanted me to forgive Al. He could have just destroyed him, and He didn't. Sure, Al died of lung cancer, but we all die. Al changed before that, and so did his ornery drug buddies, which is so much better. I don't want anybody to go to Hell, and neither does God.[2] He wants us all to go Home to Him, to know Him and be free. It's just a matter of giving up on our own particular brand of pig's food and letting Him save us.

> *It was right that we should make merry and be glad, for your brother was dead and is alive again, and was lost and is found.*
>
> Luke 15:32

---

2   Ezekiel 18:32, 33:11; 1 Timothy 2:3-4; 2 Peter 3:9

Chapter 26
Dinosaur Tracks

AJ: How did you get into studying bryozoans?
Dr. S: I dunno. Ill-spent youth.

I now study Paleozoic bryozoans because of Dr. Stillwell, but dinosaurs were always my true love. I decorate with dinosaurs. I have dinosaur bones on my mantle and dinosaur figurines lining the molding over my built-in china cabinets. You'd think I were a five-year-old boy. Dr. Stillwell loves dinosaurs too. Who can help admiring the massive *Triceratops* and T-Rex and our mutual favorite, the *Pachycephalosaurus* with its rock-hard head?

I remember my first paleontology dig as one of the loveliest, happiest times ever. We excavated duck-billed dinosaurs during the day among the sandy mushroom-shaped hoodoos of central Wyoming, giggling at Joe David's Texan sense of humor. At night, we'd get out huge telescopes and spy on the moons of Jupiter or follow Mars' path across the sky. The wide open Wyoming skies offered scenes of distant thunderstorms, and we stood out under that vastness as the lightning slashed down - or shot up from the ground and rippled along the underbellies of the far off black clouds. We all took turns cooking and doing dishes in our warm excavation community. We told jokes and stories and sang songs, and my partner-in-crime Jessica and I cooled off in the huge cow trough under the windmill.

My second dig included NASA scientists to boot. "NASA Mike" Comberiate carried around a little vial of water he'd collected from

Figure 13: "NASA Mike" Comberiate, an articulated row of Diplodocus *vertebrae, and the author as camp slave. Photo by Joe David.*

the North Pole and another vial of water he'd collected from the South Pole. How often do you meet somebody with water from both poles tucked away in his pocket?

There's something totally fun about hanging out with a bunch of NASA computer nerds on their dinosaur excavation vacation and telling them what to do. Or serving as their slave. Either way.

"Amy Joy! Is this a bone?"

"Amy Joy, what grid am I in?"

"AJ! Can you get me some water?"

"Amy Joy! Calf testicles are not 'breakfast beef!' That's messed up!"

Those NASA guys had been sitting around the campfire, innocently heating chunks of breaded 'breakfast beef' on the end of sharpened sticks before they discovered the deeper reality of the meal.

"It pains me to see you do that," my brother Baron watched them roast the baby cow bits over the fire. He'd been privy to my evil breakfast plot and seriously disapproved. But, he'd accompanied me to the farmhouse that morning and watched me bread the soft pieces and cook them up, and he'd promised to keep quiet.

It was Chandra's fault! That little ranch girl tempted me! She said offhand, "You know… they castrated the calves yesterday." How could we resist the obvious glory of feeding Rocky Mountain oysters to NASA guys when we were excavating dinosaurs just outside the Rocky Mountains? Really! Besides, I ate some too. They're chewy.

When they found out what they were eating, the scientists took the somewhat repugnant joke in good humor, but the lady who owned the ranch was furious and chomped away on little Chandra for our breakfast beef wickedness. I felt certain I'd get a verbal clobbering from Joe, but all he did was chuckle and say, "Aw. It's not a good joke unless somebody's momma gets mad."

Yeah, I love dinosaur digs. They've been some of my favorite times.

Dinosaurs died out at the end of the Cretaceous 66 million years ago when an asteroid smashed into the Yucatan Peninsula. That's what the books say. As a child, I learned about dinosaur extinction from Encyclopedia Brown in "The Case of the Cave Drawings."[1] In this short mystery, a child claimed to have found cave art that included both humans and dinosaurs, but of course they were fakes, because humans and dinosaurs never co-existed. Encyclopedia Brown knew this. He knew everything.

When he finally started the section on terrible lizards deep into that spring semester, the first thing Dr. Stillwell did was berate the "creationists out in California." He expressed unrepentant disgust with these extremists who actually thought that humans and dinosaurs had existed at the same time.

"What people need to do is find some way to lengthen the days of Creation," he said, because he believed the planet Earth was clearly older than 6,000 years.

Dr. Stillwell pulled up pictures of Hadrian's Wall to make his point. In the year A.D. 122, the Roman Emperor Hadrian started

---

1   Sobol, D. (1969). The Case of the Cave Drawings. In *Encyclopedia Brown Keeps the Peace*. New York: Nelson.

construction of this famous wall along England's border with Scotland to keep out the barbarian Picts. Local limestone was used, and after nearly 2000 years, the stone wall still stretched across the British countryside. Hadrian's Wall had fallen and been torn down in places, but those stones hadn't weathered much at all after two millennia, and limestone is not half as durable as granite or schist.

Entire mountain ranges had eroded over the history of the world. The Grand Canyon had been ripped open. If limestone Hadrian's Wall had survived one-third of the creationists' proposed timeline for Earth, it seemed unlikely there would have been time for the Appalachian Mountains to have washed away into the Atlantic or the once-jagged peaks of the Canadian Shield to wear down to piddling hills.

Dr. S. had a good point.

At the same time, I had some interesting dinosaur tracks I hadn't told him about, which added a little extra flavor to the issue.

"Are we certain that all dinosaurs died out at the end of the Cretaceous?" I asked him in his office after class. "What if some of them survived on past whatever event wiped them all out?"

He didn't seem alarmed at my unorthodox suggestion.

"How would they have survived?"

"Well, you know, the mammals stayed alive. Maybe smaller dinosaurs did too. After all, if Sir George killed a dragon, what was it he killed?"

Dragons have a ubiquitous quality in ancient art. The Vikings. The Chinese. The Persians. They all decorated with dragons.

Dr. S. shrugged and nodded. Who knew?

"Anyway, I have some interesting dinosaur tracks I think you should look at. They're casts from tracks that Joe molded."

"They aren't dinosaur prints with human tracks, are they?"

"Just look at them."

He agreed. "Okay."

I was treading into a controversial area. I didn't know the wisdom of the move, but shoot. The tracks were real and I had a personal connection to them. They weren't random tracks out there.

Figure 14: Left: a representation of a carving - the toe-marks cut down through level horizons of mud. Right: A representation of an original footprint - the mud moved under the toes before the materials hardened into stone.

For those who don't know, a number of human-like tracks have been found with dinosaur tracks along the Paluxy River outside of Glen Rose, Texas. For those readers who *do* know, just keep reading.

The dinosaur tracks along the Paluxy River are well known, and Dinosaur Valley State Park is dedicated to them. Human-ish footprints have also been found in the same location, and scores were sold during the Great Depression. A significant number were just carvings. Cheap hoaxes. Fakes are easy to recognize when sliced open and examined by cross section. The foot and toes squish into the mud in a real print, and slicing the original track in half and examining the cross section reveals this distortion of the ancient mud-turned-rock. A carving will cut down through flat layers of mud, but an original footprint will show the movement of mud under the toes.

There are definitely dinosaur tracks in the area of the Paluxy River, but some people still defend the existence of human tracks as well. They declare that the original tracks had been the real thing, but opportunistic individuals fabricated their own "human tracks" for profit when they ran out of legitimate examples.

Either way, human-like tracks have continued to be uncovered there along the Paluxy River in recent years. It's considered a non-kosher topic, and most scientists believe the prints are just misidentified dinosaur tracks. One Glen Kuban has long argued that the tracks are metatarsal dinosaur prints filled in with thin mud, giving a deceptively human-like shape to the tracks.[2] His explanation is generally regarded as the best response to the issue.

---

2   In 1986 Glen Kuban presented two papers on the Paluxy River tracks at The First International Symposium on Dinosaur Tracks and Traces in Albuquerque, NM. These papers have been published in *Dinosaur Tracks and Traces* (1989), edited by D. Gillette and M. Lockley, Cambridge University Press.

It's a touchy subject. After all, everybody knows that dinosaurs died out 66 million years ago at the end of the Cretaceous. A thin layer of iridium-rich clay shows up all over the world right at the boundary between the Cretaceous and Paleogene (formerly called the Cretaceous-Tertiary (K-T) boundary).[3] The iridium layer coincides with a giant asteroid smacking the Yucatán Peninsula, ending the Cretaceous and resulting in the deaths of the dinosaurs.[4] Boom. Dead.

I have never been to a Paluxy River dig. Old Joe, however, is an expert at making molds of fossils. He has been asked to attend digs and make molds of the tracks when they are uncovered, and the liquid latex he spreads over the tracks captures every pebble and crack. When it solidifies, a plaster "mother mold" is formed over the latex mold to ensure that the rubber keeps its shape, and a cast made from the mold can be hard to distinguish from the real thing. In 1996 and again in 1997, Joe made molds of dinosaur tracks that had human-like tracks associated with them. He didn't come in there after the fact; he attended the digs and scrubbed the rock after the limestone layer was pried off in front of dozens of people. He brought the molds back to the shop, and I made casts of them.

I have no opinion about most of the Paluxy River tracks, because I don't know about any of them. I can't say a thing for or against them. The ones that interest me are the ones that Joe molded, because I know Joe is an honest old cowboy, and I made casts from the molds he made.

A Japanese camera crew joined Joe in 1996 and filmed everything as they removed the limestone while following an *Acrocanthosaurus* trackway. When they uncovered a dinosaur track and a human-like track appeared less than a foot away in that same layer, the camera crew "ooh-ed" and "ahh-ed" and chatted furiously in Japanese. Joe said the Japanese were convinced that the creationists were right - that dinosaurs and humans coexisted after all.

---

3   Alvarez, L.et al. (1980). Extraterrestrial Cause for the Cretaceous-Tertiary Extinction. *Science*, 208(4448): 1095-1108.
4   Schulte, P. et al. (2010). The Chicxulub Asteroid Impact and Mass Extinction at the Cretaceous-Paleogene Boundary. *Science*, 327(5970): 1214-1218.

This is a picture of that track. I agree that it isn't very clear. It's mushy and puzzling. The track is about 29 cm (11.4 in) from heel to toe, and before I show people the entire track with the theropod dinosaur print near it, I like to show them this cast. I want to hear their initial reactions without any presuppositions involved, so I ask them, "What does this look like to you?" It's funny, because the first thing people always say is, "Big Foot." Or "Aliens."

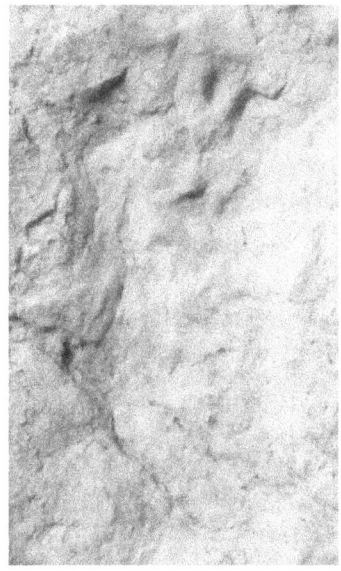

Figure 15: "Japanese" track uncovered in 1996 along the Paluxy River near Glen Rose, Texas.

I realize the very act of making a mold/cast of a print indicates that the print is special to somebody, but Sasquatches and aliens are not the most simple, likely explanation for any footprint. If I were walking down the beach and saw the tracks in the picture above, I might think, "Oh, a squirrel was running along here." I would not

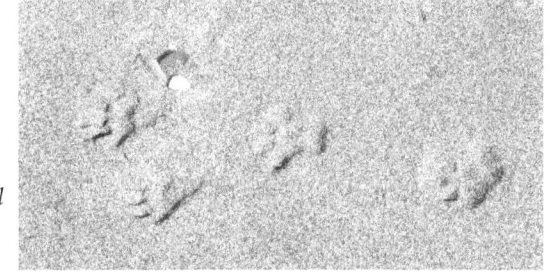

Figure 16: Presumable squirrel tracks found in the wet sand near Yellowstone Lake in Yellowstone National Park in 2014. They are not ancient.

think, "Ah! Chupacabra tracks!" Not first thing. I'd assume something terrestrial and relatively non-mythical. So, whatever made that track Joe found, I'm going to start with the supposition that it came from something terrestrial and relatively non-mythical. Below you can see the Japanese track next to the theropod track that the excavators were originally seeking as part of an *Acrocanthosaurus* trackway. I've added lines that follow the actual contours of the prints, since small pictures never do these things justice. If you want to see them in person, I have a cast hanging on the wall of my dining room.

There have been other prints in that same set of tracks, and that did interest me. An Ohio man named Hugh Miller sent Joe the cast of a print he had found years before a few feet away from where Joe found his, and it shares characteristics with Joe's track. It's just a shade shorter at 28 cm (11 in), and not as mushy. But more importantly, it has the same strong second digit impression that Joe's track has. What looks like a second toe mark sinks deep into the mud in both

Figure 17: Left: raw, untouched photograph of "Japanese" track. Right: "Japanese" track with major characteristics of both theropod and human-like tracks outlined.

Dinosaur Tracks

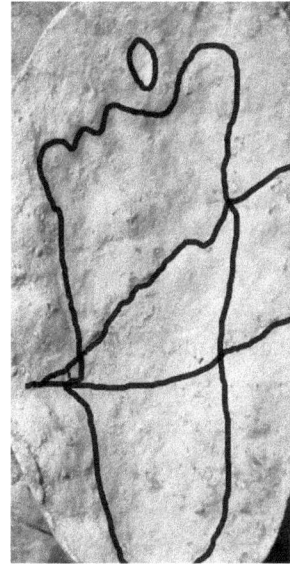

Figure 18: Track Hugh Miller found along Paluxy River near Glen Rose, Texas. The outline on the right is offered for clarity.

tracks. Joe's print is sloppy, and there is a sharp crease at the spot where the three little toes are visible in Miller's track. Strangely, the three smallest toes to the left look short in Miller's track. I'm not sure exactly what is going on here. It looks like the toes were chopped off at the knuckle - which would explain the deep second-digit mark. The toe that was left had heft. It's just an interesting pair of tracks.

The most puzzling thing to me is the clear dent on the left. Both tracks have that dent on the lower left side. It causes a crack that extends across the print to the other side in Joe's track. In Hugh Miller's track, the crack is less noticeable and it cuts straight across the foot to the other side, but it too creates a crack that veers off in a more northeasterly direction. The crack starts in the same location on both tracks. It's strange. I've stared at those two tracks for years, and I'm not sure what to make of that dent. Another toe? Another bone? A spike?

It's important to note that the prints were both made by a much lighter creature than a dinosaur, and the theropod track a foot away has clean, straight sides and shows no evidence of mud infill, contrary to Kuban's much-repeated argument. Besides, dinosaurs don't have human-like toes.

I noticed something peculiar when looking through Google Earth pictures of the Saudi Arabian Mt. Sinai site Jabal al Lawz. Among possible Exodus-era archeological remains, there are pictographs of human feet scattered about the area. Abdulaziz Alshaya posted this picture near a site called the "Rock of Horeb" by the other exploring photographers. Notice the projections sticking out of the side of the foot/sandals? Are those supposed to represent sandal straps, or did they help people walk in sand? What on earth?

I can speculate that the cracks on both sides of the prints were made by accessories on ancient footwear, but my limitations have frustrated me. I've had a deep desire to go hunt down other tracks in that same trackway along the Paluxy River, to peel up the limestone myself and collect additional variations, to get a clearer picture of the foot that made these two prints. I need more data.

Still, those two tracks didn't end it for me.

In 1997, Joe brought back another track from a July dig, and I made the cast. The three-toed dinosaur track had what appeared to

Figure 19: This photograph by عبدالعزيز الشايع *(Abdulaziz Alshaya) is posted on Google Earth at Latitude 28° 43' 28.61" N, Longitude 35° 14' 21.60" E, last accessed September 24, 2014. Used by permission, thanks to Google Translate.*

be a woman's footprint right down through the middle of it. I made the cast from Joe's latex mold and placed my foot into it. She must have been a size 9 ½, because my heel and toes fit just right. I could even follow the slipping of her ancient foot as it slid down into the most solid spot at the bottom.

A peculiar ache filled me the first time I placed my foot into that print. Somebody had given birth to the person that made it. She had grown and lived an entire life, long enough to get her feet to size 9 ½, and here I could place my foot into the form she had left in a limestone layer that had been dated to at 110 million years. It's hard to convince me that print didn't belong to a human. My foot fits it.

Of course, if it *is* a human foot, then there are two possible cursory conclusions: either humans existed 110 million years ago, or that layer has been dated incorrectly.

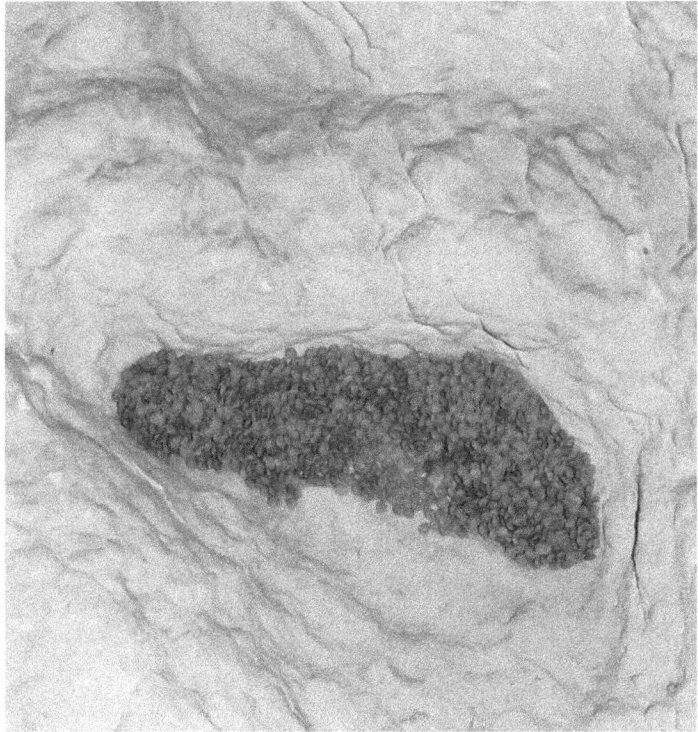

Figure 20: Track inside - and perpendicular to - an Acrocanthosaurus *track. The lentils were shaken to level to show footprint shape.*

I wanted to show these prints to Dr. Stillwell. I didn't expect them to convince him of anything. I didn't expect to make any arguments about the age of the earth. I just wanted to show them to him. Because, whatever they were, they were *real*. They had an

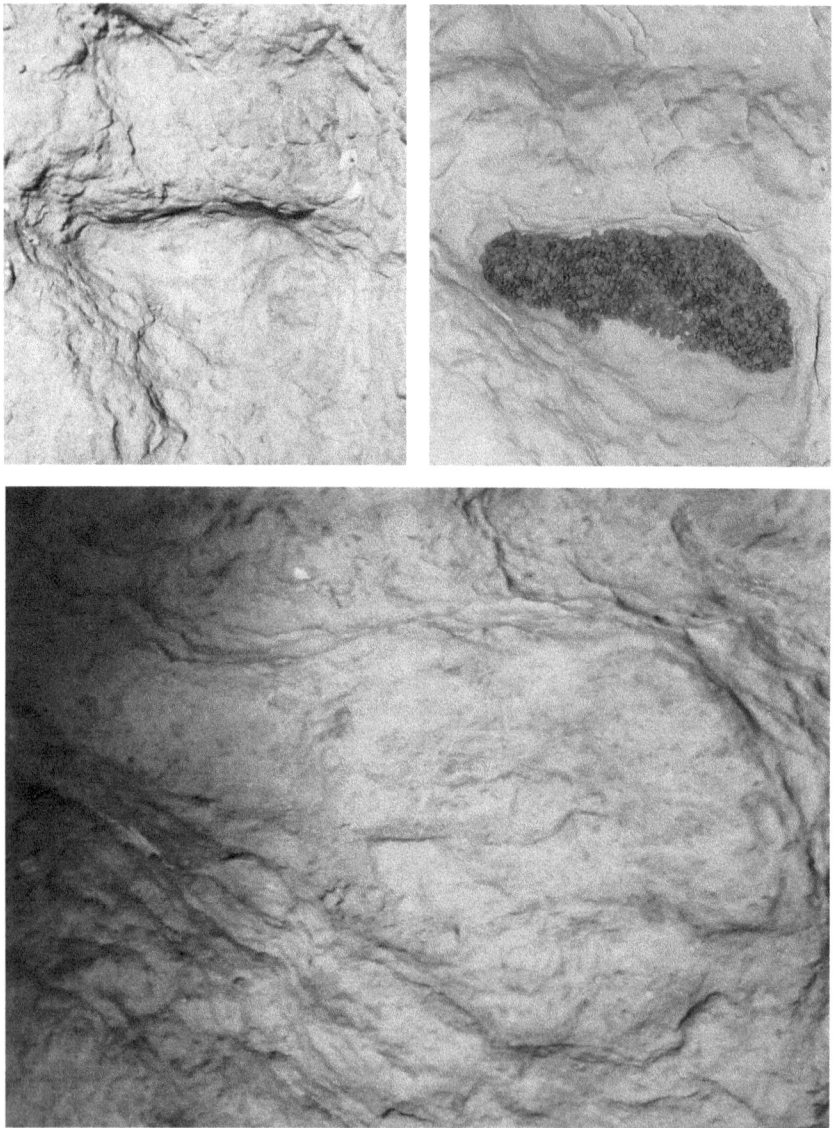

Figure 21: Top-Left: Acrocanthosaurus *track with human-like "girl" track down through the middle of it. Top-Right: same track with the "girl" portion of the track filled with lentils and shaken level. Bottom: A close-up of the empty "girl" track.*

actual correct explanation.

Mostly, I wanted him to lose his arrogant attitude.

I'm frustrated, because people see what they want to see regardless of which side of the issue they're on. It would be better if respected

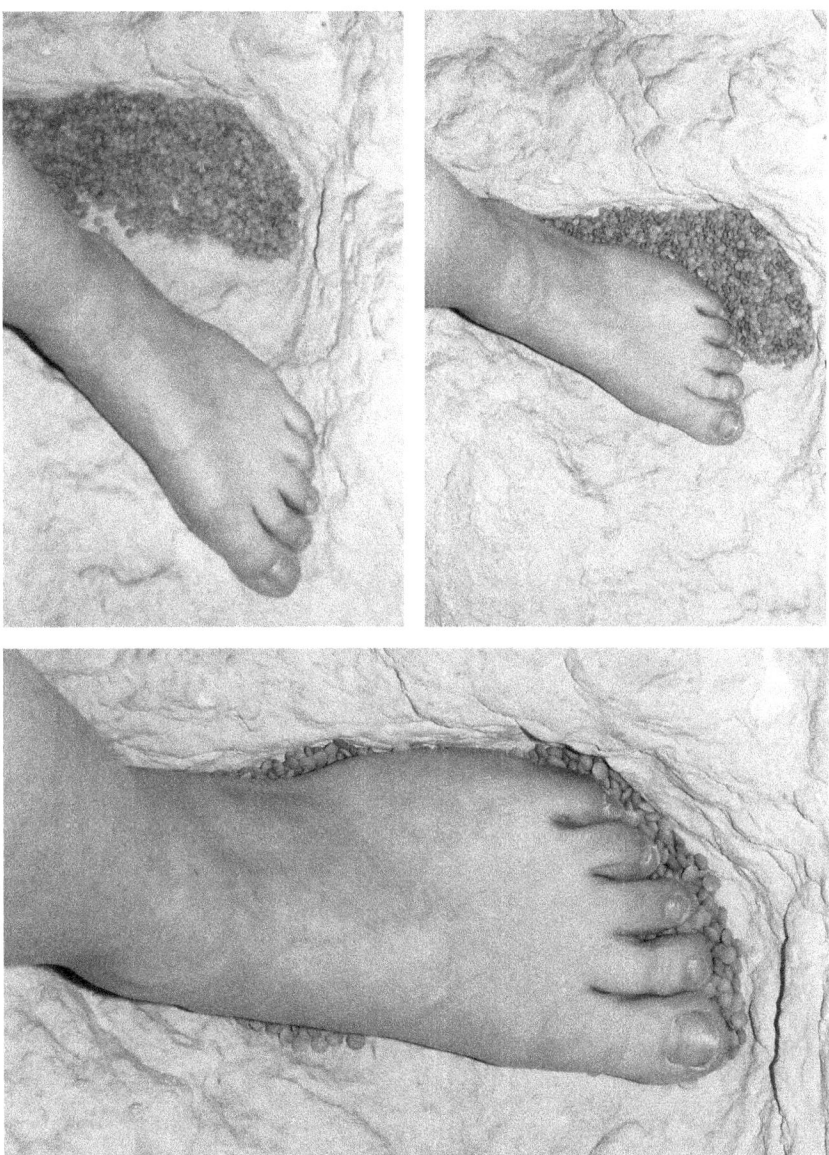

Figure 22: My left foot following the "girl" track's natural pathway down into the lowest part at the bottom. Don't judge my crooked second toe; it does its job.

groups of paleontologists catalogued every new discovery at the Paluxy River without prejudice. Honest scientists should collect a large number of prints and examine them and do 3-D analysis and put casts on display for various experts to study. Let the tracks speak for themselves.

They are just prints in rock. They are what they are.

Dr. Stillwell is not opposed to new data, but he's a geologist, and the current widely-accepted model of Earth's history doesn't have room in it for humans and dinosaurs to co-exist. The prints Joe molded were found in limestone dated to 110 million years, long before any form of human or ape or bear or raccoon was supposed to have appeared on the planet stage.

I've recently discovered how to do 3D scans, and I performed an optical scan and 3D analysis of the "girl" print. The contour marks show the sliding of her foot into the lowest spot. The big toe makes a continual pattern down to the bottom. We see an arch and even toes! She clearly put weight on the ball of her foot as she pushed off. It leaves me in awe.

Dinosaur Tracks

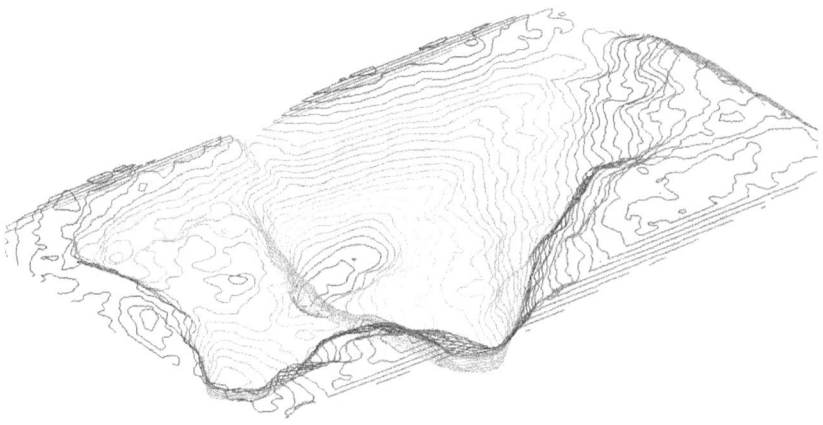

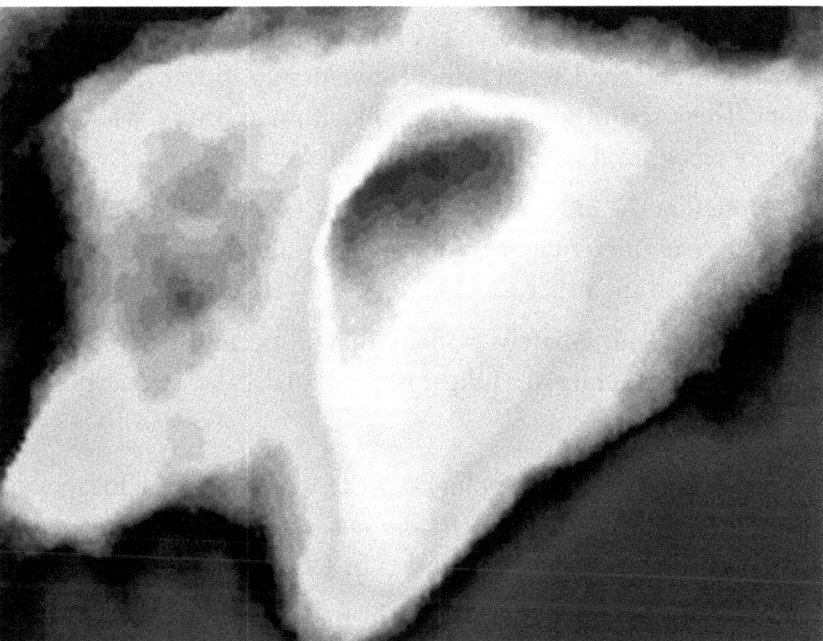

Figure 23: Two different 3-D scan renderings of the same Acrocanthosaurus *track with the "girl" print down through the middle. Top: A "topographic" representation of the prints. The ball and heel of the human-like foot are visible, as are the prints of the big toe as the foot slid down to the bottom. Bottom: Shade differences indicate depth. The ball of the foot is darker due to the deeper impression. Toe-like marks are darkest because they are the deepest. The data were captured and rendered by the author using an XBox 360 Kinect, KScan3D 1.2 scanning software and Foot Processor 1.20.3 (http://footprints.bournemouth.ac.uk/).*

Chapter 27
The Laetoli Tracks

The Paluxy tracks are not alone, of course. Photos abound of human-like footprints preserved in volcanic ash at the Laetoli site in Tanzania. The tracks are famous (in fact, I have the November 2013 issue of *Scientific American* on the table beside me, and the Laetoli prints conveniently appear right on page four). These sets of footprints look like human prints, which Mary Leakey recognized when she first found them, saying:

> Note that the longitudinal arch of the foot is well developed and resembles that of modern man, and the great toe is parallel to the other toes.[1]

> ...[I]t is immediately evident that Pliocene hominids at Laetoli had achieved a fully upright, bipedal and free-striding gait...[2]

Russell Tuttle, anthropologist at the University of Chicago and longtime editor of the *International Journal of Primatology* carefully examined the prints and concluded in 1990:

> In sum, the 3.5-million-year-old footprint traits at Laetoli site G resemble those of habitually unshod modern humans. None of their features suggest that the Laetoli hominids were less capable bipeds than we are. If the G footprints were not known to be so old, we would readily conclude that they had been made by a member of our genus, *Homo*...[3]

---

1 Leakey, M. & Hay, R. (1979). Pliocene Footprints in the Laetoli Beds at Laetoli, Northern Tanzania. *Nature* (278): 320.
2 Ibid., 323.
3 Tuttle, R. (1990). The Pitted Pattern of Laetoli Feet. *Natural History*, 99(3): 64.

Tuttle does a good job of summarizing the issue. Humans are widely believed to have appeared a little more than 2 million years ago, while the Laetoli tracks have been dated to 3.5 million years ago. Since the tracks are significantly older than humanity, according to the most popular model of human origins, it's assumed that humans could not have made them. The Laetoli footprints aren't dated to the time of the dinosaurs, obviously, but they still cause confusion because they show human-like prints in an ash layer much older than humans were supposed to be walking across the surface of the planet.

So who made the Laetoli tracks? They are widely attributed to the kin of Lucy. You remember Lucy: based on her pelvis and leg structure, she's believed to have walked upright, and she is popularly seen as a transition fossil between apes and humans. She's a famous *Australopithecine afarensis* specimen uncovered in Ethiopia (1500 miles away from Laetoli), and it's believed that Tanzanian relatives of Lucy made those 3.5 million-year-old footprints across the ash.

Anthropologists Bruce Latimer and Owen Lovejoy have produced a series of articles on the *A. afarensis* lower-limb structure, and these experts are huge fans of Lucy as a human predecessor who walked on two legs.[4] However, there's a lot about Lucy that's very chimp-like, and there's evidence she spent a lot of time moving around in trees rather than walking on the ground.[5] Her finger bones and toes were long and curved like chimpanzees, good for climbing around and gripping branches.[6] What's more, Lucy's wrists were set up for knuckle-walking.[7] She had short legs and long arms, and even if she walked upright, she didn't walk like a human.

Besides all this, we have no evidence that Lucy had human-like feet. Apes' feet are quite a bit different than human feet. Their

---

4   e.g. Latimer, B., and Lovejoy, C. (1990). The Hallucial Tarso-Metatarsal Joint in Australopithecus afarensis. *American Journal of Physical Anthropology*, 82(2): 125-133, and Latimer, B., and Lovejoy, C. (1989). The Calcaneus of Australopithecus afarensis and Its Implications for the Evolution of Bipedalism. *American Journal of Physical Anthropology*, 78(3): 369-386.

5   Ruff, C.B., Burgess, M.L., Ketcham, R.A., Kappelman, J. (2016) Limb Bone Structural Proportions and Locomotor Behavior in A.L. 288-1 ("Lucy"). *PLoS ONE* 11(11): e0166095, doi.org/10.1371/journal.pone.0166095.

6   Susman R.L., Stern, Jr. J.T., and Jungers W.L. (1984). Arboreality and Bipedality in the Hadar Hominids.*Folia Primatol*, 43:113-156.

7   Richmond, B., & Strait, D. (2000). Evidence That Humans Evolved from a Knuckle-Walking Ancestor. *Nature*, 404: 382-385.

hallux (big toe) is splayed out like the thumb on a hand. Apes' feet are flexible for gripping tree trunks and branches, and they have no arch. Lucy's toes and fingers were curved, and Tuttle has argued that if the Laetoli prints were made by *A. afarensis*, they would show evidence of Lucy's curved toes, and they don't.[8]

In March, 2010, David A. Raichlen and his team at the University of Arizona published a paper that compared the Laetoli prints to those of modern humans. Raichlen has published extensively on bipedal locomotion, and his team concluded that the Laetoli prints represent a true human-like gait and not the bent-knee, bent-hip movement of apes:

> The relative toe depths of the Laetoli prints show that, by 3.6 Ma, fully extended limb bipedal gait had evolved. Thus, our results provide the earliest unequivocal evidence of human-like bipedalism in the fossil record…
>
> …By 3.6 Ma, hominins at Laetoli, Tanzania walked with modern human-like hind limb biomechanics…[9]

Raichlen didn't argue that humans made those tracks, but the Laetoli prints are what we'd expect to see if humans walked across an ash layer in ancient Tanzania. Lucy had ape-like feet with long, curved toes and a short, waddling stride, and she might have loped around on her knuckles like apes do today. The feet and gait of her genus didn't fit those tracks.

The PBS website says something telling about the Laetoli prints in an educational page made for young people:

> The footprints also look remarkably like a human's. In fact, they looked so human-like, some scientists had a hard time believing that they were made by *Australopithecus afarensis* (Lucy's species), the only human ancestor known to have lived at the time."[10]

---

8   Lewin, R. (1983) Were Lucy's Feet Made for Walking? Science, 220:700-702.
9   Raichlen DA, et al. (2010). Laetoli Footprints Preserve Earliest Direct Evidence of Human-Like Bipedal Biomechanics. *PLoS ONE*, 5(3): e9769.
10 "A Science Odyssey: You Try It: Human Evolution: Fossil" - PBS http://www.pbs.org/wgbh/aso/tryit/evolution/footprints.html, last accessed February 21, 2014.

Nobody says the simplest, most intuitive alternative: "Maybe we've overlooked something fundamental, and we were wrong about that layer. Maybe humans walked across the ash in that location at that time."[11]

Figure 24: Southern part of hominin trackway in Laetoli site test-pit L8. Credit: Masao, F. et al.(2016) New footprints from Laetoli (Tanzania) provide evidence for marked body size variation in early hominins eLife 5:e19568 https://doi.org/10.7554/eLife.19568. Licensed under CC BY 4.0.

---

11 For a more in-depth discussion about whether Lucy's kind had human-like feet, see "The Laetoli Tracks" in the Appendix.

Chapter 28
The Dream

Dr. Stillwell and I never did get to have our conversation about geology, about erosion and the Grand Canyon and the "what ifs" on so many matters. I toted my dinosaur prints into his geology lab and stashed them under the table behind the rock saw, but I didn't show them to the geologist. He did permit me to use his rock saw to cut tiles for my new bathroom floor. Sometimes he'd banter with me on the way back to his office after class.

Really, though. The semester got busy, and Dr. Stillwell had no time for conversations. In fact, I got the distinct impression that I'd become a nuisance.

"Well, I'm going," I said one day after class.

Dr. Stillwell busily set up rocks on the lab tables in preparation for the next onslaught of students.

"Good," he said.

I decided I should make my escapes a bit faster. Nobody likes an annoying person who lingers and never goes away. You know the type, the kind that stands about grinning when you have too much to do, the kind that refuses to take the hint that you're getting undressed to take a shower and they need to get out of the bathroom? I didn't want to be that person.

One day, Dr. S. hopped out of a car at the curb just as I was crossing the street. I'd been wondering about using geology classes to earn a minor, so I turned and jaunted behind him into the science building. I could tell it wasn't a good time to ask questions. Tension tightened the air around me, and it took me only a few seconds to

The Dream

decide I didn't want to bug the professor. I carefully poked my head around the corner and saw Dr. Stillwell talking with Dr. Zenith. I let it go. I turned to head back down the hall.

Darn. Dr. S. had already seen me. He barked, "What!"

Groan. I stepped out and plodded unwillingly down the floor tiles. Dr. Stillwell stood watching me, a distinct look of impatience on his face. I didn't want to be there.

"I just wondered if I can qualify for an earth science minor," I shrugged. "I should ask Dr. Manchester. I'm sorry."

Dr. Stillwell shrugged too. "There's the Environmental Studies minor."

"I know, but I can't qualify for it because I'm a biochemistry major. The catalog says so. I wondered if I could qualify for an earth science minor. It's okay. I'll talk to Dr. Manchester. He's my advisor."

I started to leave, but first I asked, "Hey. Was that your cute wife who dropped you off just now?"

Dr. Stillwell's whole countenance changed. He brightened. He stood straighter and his face stretched in a grin.

"You guys eat lunch together?"

He nodded.

"Well, that's cool," I approved cheerily. "Anyway, thanks."

Crudola. I had become *that* annoying student. And not just a student. A female. A widow. Sigh. That's not what I'd intended at all.

I needed to seriously back up a bit. After that brief exchange in the hall, I made a conscious effort to retreat. I left the lab right after class every day and didn't linger to talk geology mysteries. I didn't interrupt Dr. Stillwell at his office. I stopped sleeping on his front porch. Hahaha. Just kidding. I didn't stop sleeping on his front porch.[1]

Then, one night at the end of the semester, I asked God something I'd never asked before. Looking back, I think it's strange that I asked it. Dr. S. and I weren't even talking at that point. It was the end of the semester. He was busy. I was busy…

---

[1] Somebody… somebody is going to seriously think I go around sleeping on people's porches. People. Please. Porch floors are hard, and it gets cold and sometimes the rain comes in at an angle and… No.

I don't know. It bothered me that Dr. Stillwell had a disdain for people of faith. He clearly despised those of us who believed in a deity, as though we were stuck in the Dark Ages. I wanted him to know that God was alive and well, that He was still doing things in our nanochip and plastic wrap world.

So I asked God for a dream about Dr. Stillwell.

I wanted God to show me something about Dr. Stillwell that nobody else knew as evidence that God was real and working in my life. I had never made that kind of request before, and it was something I wouldn't normally dare ask. On that particular evening I felt like it was okay, that God heard me and would do it.

We have to be careful about dreams. Our brains spit out all kinds of crazy things at night, and we shouldn't read too much into their random pictures. If somebody asked me, "I had this weird dream last night. What does it mean?" I would probably answer, "It means that you slept long enough to have a lot of REM time in the morning. That's what it means." In other words, "It doesn't mean anything."

On the other hand, God gave people dreams throughout the Bible. He spoke face to face with Moses, but other prophets were limited to dreams and visions.[2] It's all over. God gave dreams to Joseph the son of Jacob, and He gave dreams to Joseph the earthly father of Jesus.[3] God gave dreams to men who didn't even serve him, like Abimelech and Nebuchadnezzar.[4] Joel 2:28-29 tells us that in the last days, prophecies and dreams and visions will be common:

> *And it shall come to pass afterward That I will pour out My Spirit on all flesh; Your sons and your daughters shall prophesy, Your old men shall dream dreams, Your young men shall see visions. And also on My menservants and on My maidservants I will pour out My Spirit in those days...*

Sons and daughters, male and female servants.

So, I asked for a dream.

As I said, I felt good about it. Of course, I wanted the dream

---

2 Numbers 12:6-8.
3 Genesis 37:5-20; 2:13,19
4 E.g. Genesis 20, Daniel 2

to be about Dr. Stillwell's childhood, about his little red wagon or some important event that mattered to him - a view into something only he knew about. That's what I wanted. I wanted to perform a parlor trick. I've discovered that God doesn't perform parlor tricks. It turned out there were bigger issues at stake than red wagons.

I had a dream, and it wasn't about Dr. Stillwell's childhood at all. It was about his heart toward me.

I felt the emotion before anything else. It was this warm, comfortable, open, friendly feeling. Even as I was dreaming it, I was trying to describe that emotion. Warm. Comfortable. Open. Relaxed. As I was feeling this, I looked over and saw Dr. Stillwell standing there. He was somewhat embarrassed in the dream, because it was easy for something comfortable and open like that to look bad, for people to take it the wrong way. But, it was innocent. He knew it and I knew it.

The emotion did not include "affectionate." I realized that wasn't the right word. That word didn't fit. It was warm and open and friendly and comfortable, like fluffy down pillows and slippers. Like floppy-eared puppies. Like Saturday morning coffee. It made me think that we were like childhood friends, or like a father and his daughter. That kind of relationship.

When I started looking around in the dream and I saw him standing there, though, I flipped out. I didn't care that it was warm and comfortable and totally innocent. As soon as I sat up and my eyes landed on him, I jumped up and walked out of the room. It looked bad his being so close and comfortable with a co-ed, and I wasn't staying. He followed me out, saying, "Would you relax? Would you please relax?" as I headed down the hall.

I woke up about five seconds later, and I thought, "What in the *world* do I do with that? I can't tell him that!" If I told it to him, he'd think it was just my unconscious mind working, and that would be completely awkward. It wasn't about me, it was about *him*. Which was unexpected, because he thought I was annoying and didn't have time for me.

"I can't tell him that!" I told Matthew Caerphilly. "Good grief, what am I supposed to do with a dream like that?"

"Maybe you're not supposed to tell him," Matthew said. "Maybe that's not the purpose of it."

So, I didn't tell Dr. Stillwell about the dream. Not for another year. It was just a sort of bizarre, somewhat embarrassing thing to describe to one's geology professor. I mean, really. So, I didn't.

The very morning after the dream - that very morning - I took my final exam in Historical Geology. I used the entire two hours to finish - a lone, paranoid overachiever scribbling away long after her peers had handed in their essay questions. Stretching out my cramped writing hand, I dropped my papers onto the pile of other exams and saluted Dr. S. as I passed him. Then, I headed out the door and down the hall.

I got about 15 feet away. That's it, because Dr. Stillwell surprised me by stepping out his lab door and calling out after me. I don't remember what he said, but it was the sort of question that had absolutely nothing to do with anything. He could have asked, "Do you tie your shoes with bunny ears or with the loop and swoop method?" That would have been just as relevant.

As soon as he shouted, I stopped and turned around, the hall rotating around me. Time slowed. I saw the posters on the walls and the door to another lab and the tiles on the floor. The fluids in my inner ear swooshed. "It's true," I thought as I made my way back toward him. "That dream is true."

I didn't know whether he had the warm, comfortable friendship feeling about me right then, or whether it would come. But, I knew he had just stopped me from leaving and had asked me a pointless question so that I would have to go back.

Maybe it wasn't the fact that I'd had the dream that mattered, like Matthew said. Maybe it was just that lattes and gingersnaps sort of camaraderie... maybe that's what God had in store for us.

I did finally tell Dr. Stillwell about the dream, nearly a year later when it wasn't quite as awkward. In response, he confided with honesty, "There are just very few people with whom you can feel truly comfortable."

Chapter 29
Not My Business

I didn't get to show Dr. Stillwell my dinosaur tracks that semester at all. I didn't get to talk to him about them. I didn't get to be around when he finally had a chance to peek at them. He was terribly busy, remember, and every time we might have been able to look at the casts something would go wrong. He wouldn't let us into the lab until the very last minute before class started. His wife's car broke down. My car broke down.

I stopped into Dr. Stillwell's office after all exams were over, the very very very last day that I would be at school before flying to Seattle to start summer break. Dr. S. was going to let me use the rock saw to cut the marble threshold for my remodeled bathroom. The professor was going to take a moment to look at those tracks with me.

I popped my head into his office that Friday morning. "Will you be around about 11?"

He said, "I'll be here!"

But he wasn't. He wasn't there at 11:00. He wasn't there the rest of the day.

That's when I received another verse about Dr. Stillwell.

I sat in the hallway outside Dr. Stillwell's locked office door, wondering whether he would ever come back. My marble threshold was sitting in his lab. My dinosaur tracks were sitting in his lab. I was getting on an airplane the next day.

Out in the hall, God gave me a verse that told me, "He's my business, not yours." I sat on the floor in the hall, pulled out my Bible and opened it up. There it was in black and white. I know it's lame not to give you the verse itself, but to my frustration I didn't write down the exact words and I haven't been able to find it again, but that's what the verse said.

God was in charge of Dr. Stillwell. It wasn't my job to change his mind about anything. It wasn't my job to show him dinosaur tracks with human-like tracks squished in and around them or to counter his arrogance about geology or religion or tell him dreams or anything. That wasn't my job.

"Okay, Lord."

And so, the dinosaur tracks and my marble threshold sat in the geology lab all summer long. A time would come when Dr. Stillwell and I looked at those casts together, face to face, but not that semester.

I'm certain Matthew Caerphilly was right about my dream. If Dr. Stillwell were ever to know the God of the Universe still worked in human lives, it would be something that the God of the Universe would have to accomplish. God would have to show Himself to the man. It wasn't my responsibility, and that was fine with me.

Chapter 30

Randy

Dr. Stillwell is a geologist who has rejected God. He's not a bad guy. He's a good scientist and one of my best friends, but I think he's like a lot of scientists. They don't reject God because of fossils or anything they see under a microscope. They reject God for personal reasons, reasons close to their own hearts.

I've talked to Dr. Stillwell many times during the past few years, and his disbelief never has anything to do with rocks. It always returns to the age-old issue of pain. What about Grandma? Why is there pain in the world?

I can ask that question myself. Why did my husband suffer so long and die so young? Why did he have to leave me and the children behind?

I don't know. But, as I end this volume, I need to tell Randy's story. It deserves to be told.

When Randy was a little boy, he got straight A's. He loved music and played the trumpet. One great uncle took him under his wing and taught Randy the Bible, and that bright young man spent every week at his Lutheran youth group. He ran cross country and won races. He loved to run, and he carefully documented the number of miles he covered each day.

Native American ancestors gave Randy his black hair and dark eyes, but one tiny Cherokee great grandmother stole his height. Fourteen-year-old Randy was small for his age and had a gentle, young face, and he attracted the attention of a local pedophile. It didn't matter how much Randy wanted to get out of it, he couldn't escape Dice. He was forced to return time and again, and it became clear he wasn't the only one. Dice had a whole gang of young men he controlled.

Two years into the relationship, Dice threatened a child who didn't even exist yet. He told Randy, "I'm going to get your son. You wait. One day I'll get your son too."

That did it. Randy grabbed Dice around the neck and jacked him against the wall. He'd grown strong, and Dice struggled futilely against his 16-year-old victim's hands. The staircase loomed beside them and Randy could have thrown Dice down the stairs - maybe killed him. Instead, Randy lowered his tormenter to the floor and left that house for the last time.

Still. Those two years had done their damage. Randy hated himself, and shame chewed away inside him. His grades plummeted. He plunged into drugs and alcohol and waded through a stream of girlfriends. He managed the McDonald's in town but found it was more lucrative to run drugs up and down the coast from Miami.

"I got into bar fights, and I scared the other men because I just didn't care if I lived or died," he once told me.

Years later, Randy and I took a stroll along the Spokane River in Post Falls, Idaho. I knew all these things about him, because he was always honest about them. We'd become good friends, and we'd started talking about marriage. As we walked along the rocks above the river, I asked offhand, "Randy, how many women have you been with anyway?"

He cringed at the question. "Do we have to talk about that?"

We stood out on a rock outcrop overlooking the Post Falls waterfall. "Randy. How many?" I was planning to marry him. I wanted to know.

He took a deep breath. "Oh I kept track at one point." He thought about it. "Um… less than 100. No. Less than 80."

That stopped my breath.

Randy had always been open about the messiness of his past. He'd never tried to hide anything, but in my young naiveté I'd been thinking maybe 10 ladies, so his "less than 100" really punched me in the chest. I wanted to breathe and think about that, take time to process it. All those women. Dozens and dozens and dozens of partners before me. I'd been okay with 10 women, but there wasn't any real difference between 10 and 100, because every one of those women was one too many. His admission drove home the heart of the thing. Those women were all somebody other than *me*, and I didn't want to marry somebody else's man. The whole thing was impossible!

I'd come out of a difficult childhood and had a hard time opening my heart to anybody. I hadn't jumped into the arms of this mountain man from West Virginia. I enjoyed his friendship a great deal, but that's all I'd wanted from him. Yet, God had carefully ushered me into a closer relationship with Randy, and God had spent patient time convincing me He was behind it. As Randy and I walked back to his house, I knew I was missing something. If this relationship was God's idea, then it was okay. What was I missing?

I looked at my future husband and realized what it was.

"Randy," I said earnestly, "you have to be a brand new person. That's all there is to it. You have to be *brand new*."

"Well, I am," he shrugged.

It was true. He was a different person. He'd left that angry, broken young man behind.

Randy had been a wreck for a lot of years. He'd destroyed his first marriage, and when he married his second wife, he'd tried to clean himself up. He had children and wanted to be a good dad. He went to Alcoholics Anonymous for eight months. "But, I was miserable the whole time." He drank heavily because he hated himself, because he felt worthless, because he was full of anger and shame. When he wasn't drinking, he felt the full force of all those painful emotions.

"I was like a dog any woman could lead around," he told me about himself later. "I didn't care about myself. It's only the mercy of God that I didn't kill anybody."

On New Year's Day soon after his firstborn Brandon turned seven, Randy spun out on the ice and smashed into a rock wall. He'd gone to pick up four-year-old Amber, and she rode in the passenger seat next to him. The airbag deployed when he bounced off a tree stump, but it deflated by the time he hit the hillside. He smashed his face and nearly snapped his already-damaged neck. He looked over and saw a bubble around little Amber. She swears he put his arm over to protect her. Either way, she escaped unscratched.

In the hospital, the doctors murmured over the fact that Randy had survived the wreck. They thought his neck injuries should have killed him. At that point, Randy said, "Okay, Lord. You just spared my life. I've made a complete mess of it, so whatever You saved me for, here's my life. You can have it."

I met Randy four years later, when he loved God with all his heart. I never knew the Randy who ran drugs up and down from Miami, who suffered in misery without his whiskey. Plenty of other people told me about it, but I never met that guy. The Randy I knew had a glass of wine on Thanksgiving and Christmas, but it was years before I even saw him take an Advil. He talked openly about his self-destructive youth, but the people in our group didn't make a big deal about it; we didn't see any traces left.

When I met Randy, he loved everybody he met. In the days before cell phones, he'd always stop for people broken down beside the road. He readily helped anybody who needed help, and when he prayed for his guys at work, God answered the prayers. He came home from work one day and said, "Well, Jesse got his foot run over by a fork lift today, so I prayed for him. Then we put him in the car and I took him to the emergency room. By the time we got there, his foot was completely healed. There were no broken bones on the X-ray, and they couldn't figure out where all the blood in his sock had come from."

Our friend Kelli Thomas limped in constant pain from old injuries to her knee, injuries that several surgeries hadn't soothed. I prayed for her, and she wasn't healed.

Yep. I prayed for Kelli Thomas. She wasn't healed.

A few months later, Randy and Jim Mader prayed for Kelli. "Wow, Kelli," Jim said. "You're still limping? C'mon, sit down here and we'll pray for you." Boom. She was healed. I stood there, watching them. As they prayed, Kelli began to cry. She later said she felt a burning in her knee, then all the pain went away. She hugged her dear husband and walked out with a new knee.

"I still limp from habit," she told me a few weeks later. "I'm so used to limping. But, I don't need to. It's completely fine."

Randy (and Jim) prayed for people and they were healed. I don't seem to have that gift. I wish I did, but it's okay. Kelli was healed, and that was the important thing.

Randy was a blessing to me in a lot of ways. I still couldn't hear God well in those days, but Randy could. I'll admit that's one reason I liked being around him; God did things in him.

Randy gave his life over to God after the car accident, but it was after he got hit by a front end loader that his life fully changed. Two years after Randy's spinout on the ice, a fellow worker ran into him with a front end loader. The accident herniated discs in Randy's back, and he spent his recovery time in bed reading the Bible. He poured over its pages, absorbing their words like life. On a rainy day that summer, he stood out on his front porch, and the Holy Spirit came upon him and overwhelmed him. God filled Randy with joy and healed that crushed and battered heart of his.

"I stepped down into the grass and danced in the rain," Randy told me years later. "I hadn't felt loved as long as I could remember. There in the rain, God filled me with His love, and it changed everything. I'd hated myself for so long, and He washed away all the shame and guilt, and He filled me up inside. When I knew God's love for me, then I was able to love other people."

Randy's marriage still fell apart. His wife said, "I know who you really are, and you're not some holy man." I'm sure his new-believer enthusiasm drove her crazy. She was used to being the strong one, the good one. She'd put up with him all the years he was a mess, but she didn't want a life with the new Randy. She forced him out and got a new man. She wouldn't let him see the girls. She had the money. Her mother was in charge of child social services in town. Randy felt helpless.

Randy came out to Idaho to earn money and got stuck there.

He met me, and as much as we wanted him to see his daughters, it was a struggle for years.

But. But, he was whole. He was healthy. He was strong and worked harder than any two men. He fell in love with me, and God gave us three beautiful children. Amber came out and lived with us. We worked with the youth group, and he became a surrogate father to other hurting young people.

Four years into our marriage, we moved back to West Virginia to be near his kids. Still, Randy was only able to see his oldest children. He didn't get to see his younger daughters regularly, no matter what he did. Once a year - that's all their mother allowed. The girls wanted to know him, but their mother was hard and unwilling to let him in.

Then, Randy's back and neck started hurting him again, and his old doctors put him back on pain meds. Hepatitis C from years before hadn't ravaged his liver, but he feared it would catch up one day, so he went through an intense treatment regimen. The medication eradicated the virus but all but destroyed his immune system. He was falling apart.

A few months before he died, I stood in the kitchen at the cabin arguing in my head at Randy's second wife. Her daughters wanted to see their father, and she made it painful for everybody. She screamed at Randy any time he called, reminding everybody of his sins from years before. It was horrible and wrong, and she wasn't protecting anybody. All she was doing was making her children lose their father. I yelled at her in my head, frustrated and angry at her as usual. And then, like usual, I stopped yelling at her and started praying for her. The yelling certainly didn't do any good, but the praying could.

As I prayed, God spoke to my spirit as He'd done many times before. He spoke simply and directly, and I knew it was Him. He said, "What if the girls never have a real relationship with their dad if it means their mom will be saved. Would you go for that?"

That's how He phrased it, too. He said, "Would you go for that?"

That stopped me. "Well. Of course. But… does it have to be that way? He can't stop trying to be in their lives. He can't just blow them off."

It made no sense. Randy always treated the girls' mother with patient decency, but he couldn't stop trying. The girls needed to know their father loved them, it was important for their hearts to have a good relationship with him. I didn't understand, but I decided to let God handle it.

Randy got double pneumonia in May of 2009. It went septic and he almost died. That July, the girls' mother let him come visit, and he had a wonderful time with his youngest daughters. There were hugs and tears, laughter and forgiveness. He remembered that time with joy every time he'd talk about it.

He saw all his children that summer. The kids all enjoyed him, and none of them knew it was the last time they'd see him. When he died in September… well, that made a lot more sense than his giving up on any of them.

During those years when he was a mess, Randy did horrible things to other people. He cheated on his wives. He sold drugs. He beat people up. He broke hearts. God let him do those things; He didn't stop him.

But, I can testify that God rescued Randy and healed his heart. I'm certain Randy is in Heaven, enjoying the Savior who loves him and gave him back his life, and I can testify that God really did make a brand new man out of that broken human being. Randy became a blessing to a lot of people, including me. It was Randy who helped me really feel God's love - really feel it. And it was after I started having confidence in God's love for me that I found my prayers answered all the time.

What if Randy had remained hard? He didn't have to say, "Okay God. Here's my life." He could have ended up like so many who die angry and bitter and hated. As soon as he handed his life over, God began to pick up the broken pieces, and even in Randy's death, I felt the purpose of it.

"Would you go for that?" God had asked me.

I agreed to it. I didn't know Randy was going to die, but I did agree to it. As I look back over the years, I see how many other people were involved in that agreement. My stepdaughters and their mother were involved, yes. Randy's unborn grandchildren were involved. Dr. Stillwell was involved too, because I'd never have taken his 8 o'clock classes if Randy were alive. I wouldn't have become Dr. Stillwell's friend. I wouldn't have written this book. Which means that you were involved too.

God loves us, and He's much smarter than we are. He sees far beyond what we see.

We think there's a giant chasm between religious belief and science, but that's not really so. The evidence will point to what's true if we have enough of it. The real chasm is always between us and knowing God. That chasm is real. Still, we haven't been abandoned. Not for a minute.

When little Paul Stillwell kicked and screamed and refused to go to his catechism classes, I imagine that God saw that child and loved him. Life has done a lot to Dr. Stillwell, but in the end he hasn't changed. In the depths of his soul, he's still a little boy who loves rocks, walking along the edge of the chasm. He looks like he's alone and in danger, but I think God has been there all along, ready to catch him up and carry him across.

This story is just started - if you'll hear it out. We have to tell about my great larks with Dr. Zenith the astrophysicist. We must get to the ninjas. Most importantly, God started doing things in my life with Dr. Paul Denali Stillwell, atheist geologist. God started doing things, and I cannot possibly fit it all into one book.

Figure 25: The author at the Grand Canyon in 2010. Photo by Jordan Rahert.

APPENDIX

1) A Worldwide Flood?.................. pg. 179
2) Who Wrote Genesis? .... pg. 190
3) What is the *Archaeopteryx*?........ pg. 208
4) The Laetoli Tracks .... pg. 224

A Worldwide Flood?

In December of 1989, a fellow named Sherm Byers found the nearly complete skeleton of a mastodon on his golf course in Licking County, Ohio. It's hard to describe the significance of Sherm's extinct elephant. It had been butchered by early Americans thousands of years ago, and knife marks showed on its bones. The contents of its stomach told paleobiologists that the big guy ate swamp grass and weeds instead of the spruce twigs they'd expected. Most importantly, mastodon-aged bacteria were still locked in the region of its bowels.[1]

Lead excavator Brad Lepper of the Ohio Historical Society said, "It was the first time that microbiologists revived the bacteria so that we had a living creature from the ice age. We had a potential window, frozen in time. We were in the (*The Guinness Book of World Records*) for having the oldest living thing for a while. But then someone found bacteria in bees' guts dating back to billions of years old."[2][3]

That find didn't hurt Sherm at all. His Burning Tree Golf Course received instant publicity as the site of the famous Burning

---
1   Wilford, J. (1991, May 4). Mastodon Yields Living Organisms. *The New York Times*. Retrieved March 7, 2015 from http://www.nytimes.com/1991/05/04/us/mastodon-yields-living-organisms.html
2   Whyde, L. (2008, June 16). Mastodon Skeleton Discovery in Heath Changed Local Man's Life. *Newark Advocate*.
3   Leppers "billions" is actually bee bacteria dated to 25-30 million years ago. Most life on earth is supposed to have evolved in the past 541 million years since the Cambrian started, and bees didn't appear in the geologic column until the Cretaceous. The bee bacterium, in turn, has been trumped by 250 million-year-old bacteria trapped in halite salt crystals: Vreeland, R. (2000). Isolation of a 250 Million-Year-Old Halotolerant Bacterium from a Primary Salt Crystal. *Nature*, 407: 897-900.

Tree Mastodon. Joe David made the mold and first casts of Sherm's mastodon, and then Sherm sold the original skeleton to a Japanese museum for $600,000.

Many years later, I paid a visit to Mr. Byers on his golf course. My brother Lex dropped me off, and I wandered across the yard, trying to figure out where Sherm kept the front door to his enormous house. It was after dark, and I couldn't see anything but windows and more windows, so I finally started backing up to get a better perspective.

And I fell into Sherm's pool.

The pool did have a big plastic cover over it, so I splashed around up to my knees before clambering back out. Lex still parked in the driveway, shaking his head.

Half an hour later, I perched on Sherm's couch in dry clothes, thanks to his wife. I expected to talk about the mastodon, the golf course, and the work we would be doing over the next few days. Instead, Sherm went off about Noah's Flood. What on earth! He ranted about how all the animals could never have fit on the ark, how they could never have survived the trip, and how the entire thing would have been impossible. I didn't know what to do. I was young and a guest in his home, and I hadn't initiated the outburst.

It makes a good story, my falling into Sherm Byers' pool. And it's a true one. A worldwide Flood sent to destroy the depraved human race makes a powerful story, but does it have any basis in reality? Could Noah have built a seaworthy vessel, into which representatives of all land animals fit and survived? Are the physics and geology involved even a little possible?

I can't do justice to those questions here.

However, I can make some notes. I do know that stories about a global flood are found all over the world. Aside from the many flood legends, it's mentioned as a real event in the Sumerian King List. The Bible treats the Flood as a true story, and it treats Noah as a real person throughout the whole thing - from Isaiah and Ezekiel to Peter and the writer of Hebrews.[4] Jesus describes Noah as a historical person and uses him as a warning about the future:

---

4   Isaiah 54:9; Ezekiel 14:14-20; Hebrews 11:7; 1 Peter 3:20; 2 Peter 2:5

But as the days of Noah were, so also will the coming of the Son of Man be. For as in the days before the flood, they were eating and drinking, marrying and giving in marriage, until the day that Noah entered the ark, and did not know until the flood came and took them all away, so also will the coming of the Son of Man be.

Matthew 24:37-39

I also note that the Bible's account in Genesis gives *dates*. The flood story that Utnapishtim tells in the Epic of Gilgamesh offers a length of time it took to build the boat and wait out the rain, but Noah's account gives the actual month and day that certain events took place. As a journalist, I regard this detail as significant.

In 1961, Henry Morris and John Whitcomb published one of the first creationist books, *The Genesis Flood*, as an effort to look at the Flood through geological eyes. R.A. Moore's 1983 article, "The Impossible Voyage of Noah's Ark," rejects the Bible's story as ridiculous, and in his 1995 book *Foundation, Fall and Flood*, Glenn Morton argues the Flood was a local event in the Middle East. In 1996, John Woodmorappe addressed questions like, "Would the animals all fit?" in *Noah's Ark: A Feasibility Study*…

People have been arguing about this thing for a long time.

A quick trip to TalkOrigins.org offers Mark Isaak's, "Problems with a Global Flood."[5] In his article, Isaak asks a wide range of questions from the standard, "Were dinosaurs and other extinct animals on the ark?" to more serious geological and physical questions like, "How could a flood deposit layered fossil forests?" or, "How does a global flood explain angular unconformities?" Jonathan Sarfati offers responses to many of Isaak's concerns,[6] and Kurt Wise argues that a global flood is the best explanation for the

---

5   Isaak, M. (1998). Problems with a Global Flood 2nd Edition. Retrieved February 24, 2015, from http://www.talkorigins.org/faqs/faq-noahs-ark.html
6   Sarfati, J. (2015). How Did all the Animals Fit on Noah's Ark? *Creation Club* Retrieved July 16, 2024 from https://thecreationclub.com/how-did-all-the-animals-fit-on-noahs-ark-jonathan-sarfati/

vast sediment layers we find.[7] We really do find layers of sediment that cover entire continents.

I reject a dismissal of the Flood, but the God I serve seems very practical. He lets things play out. He lets the chips fall where the laws of physics demand them to fall. If the Flood took place globally, then it would have left behind serious telltale signs. It would be recorded in the rocks.

Dr. Stillwell and Dr. Zenith laughed at the Bible's story, because they believed that the existence of a firmament would have increased Earth's atmospheric pressure and cooked everybody. I thought perhaps the amount of pressure didn't have to be that great, but they were right about the effect of a canopy. Even creationists have agreed that a water vapor canopy couldn't hold water more than about 2.0 meters deep without making Earth's surface temperature uncomfortably hot - even if the solar constant, albedo (reflectivity) of the atmosphere, the solar zenith angle and cirrus cloud thickness were optimal to keep surface temperatures low.[8]

What does that mean for the Bible's description of the *raqiya*, the "firmament?" I don't know. We don't even know what the firmament actually was, and I don't assume it's just the stuff of ancient imaginations. Genesis says that on the 17th day of the second month all the fountains of the great deep were broken up and the windows of heaven were opened. The fountains are mentioned first, so it's reasonable to suggest that a large amount of water would have come from under the Earth's crust. Still, what would that have looked like? What geologic processes could have produced eruptions of water around the world without boiling everybody on board the ark? Would the land masses sink below sea level in one place while magma shoved it up in another? Did fissures open across the Earth's crust, hurling the continents around like bumper boats? Can we make realistic predictions about what we should find in the rocks or do we leave out important pieces because we just don't appreciate

---

7 Wise, Kurt (2017). 90 Minutes of Geological Evidence for Noah's Flood. Retrieved July 16, 2024 from https://www.youtube.com/watch?v=882fmumdm9A
8 Vardiman, L., & Bousselot, K. (1998, August). Sensitivity Studies On Vapor Canopy Temperature Profiles. Retrieved March 3, 2015, from http://www.icr.org/research/index/researchp_lv_r05/

all the variables?

The best person I know to address these issues is a gentleman named John R. Baumgardner, who earned his PhD in geophysics from UCLA and went on to work for Los Alamos National Laboratory. He developed the computer program TERRA to model plate tectonics and the forces that drive volcanic eruptions and earthquakes,[9] and he spent multiple decades developing a theory of Flood geology.[10][11] His articles on catastrophic plate tectonics are available online, and if you want some answers, you should read them. After 25 years of research, he summarized his position at his site globalflood.org:

> My conclusion is that the Flood was one facet of a larger global-scale tectonic cataclysm. A key aspect of this catastrophe was the rapid sinking, in conveyor belt fashion, of the pre-Flood ocean tectonic plates into the earth's interior. The energy required for the process was derived from the earth's gravity acting on the excess weight of these cold ocean plates relative to the hotter and less dense mantle rock into which they slid.

More recently, Kurt Wise produced a series of videos available under "Is Genesis History?" on YouTube. Wise is a global flood proponent who earned his PhD at Harvard under Stephen Jay Gould, and he's addressed the geophysics involved.

I want to go in another direction in this short appendix. The night I fell into Sherm Byer's pool, he yelled at me that the story of Noah's Ark was not feasible, and that's where a lot of people start. For instance, Isaak's first complaint at TalkOrigins.com is that a wooden ark the size of Noah's might not be sea worthy, because large wooden

---

9   Tackley, P., & Baumgardner, J. e al (1985). Three-Dimensional Spherical Simulations of Convection in Earth's Mantle and Core Using Massively-Parallel Computers. *Journal of Statistical Physics*, 39: 501-511.
10 See Baumgardner, J. (2003). Catastrophic Plate Tectonics: The Physics Behind the Genesis Flood. In R. Ivey (Ed.), *Proceedings of the Fifth International Conference on Creationism*. (pp. 113-126). Pittsburgh, PA: Creation Science Fellowship. Retrieved April 28, 2015 from http://static.icr.org/i/pdf/technical/Catastrophic-Plate-Tectonics-The-Physics.pdf
11 Also see Austin, S., Baumgardner, J. et al (1994). Catastrophic Plate Tectonics: A Global Flood Model of Earth History. In R. Walsh, (Ed), *Proceedings of the Third International Conference on Creationism*. (pp. 609-621). Pittsburgh, PA: Creation Science Fellowship. Retrieved April 28, 2015 from http://static.icr.org/i/pdf/technical/Catastrophic-PlateI-Tectonics-A-Global-Flood-Model.pdf

boats today leak badly. This seems to be a summary of an argument in Moore's 1983 article. I want to criticize Isaak a little, because he highlights a serious widespread problem we have in the 21st century: our heads are too fat. We don't know how the Saksaywaman fortress in Peru was engineered, and we work to determine how the pyramids were built, and we argue over how the Easter Island statutes were hauled from miles away and set up. We think we're the smartest humans that have ever lived. Yet, the pyramids, Saksaywaman, and Easter Island each give us evidence that the peoples in our past were bright too. Very bright. They possessed technology and engineering skills we can't imagine.

Baalbek is a town in the Beqaa Valley of Lebanon, where the Romans built a massive sanctuary to their gods. The foundations of the sanctuary appear to be much older than the Roman structure constructed on top, and the western wall includes the famous trilithon, a row of three stones that weigh approximately 800 tons each. That is, 1,600,000 pounds each. Each of these three stones are 70 feet long and 14 feet high (Fig. 26). Enormous. The stones were cut from a quarry (Fig. 27), moved about a kilometer to the temple site, and neatly raised into place on stones that weighed 300 tons each (also very large). There has been a great deal of speculation about how these monstrously huge limestone blocks were cut and moved there. The Romans used much smaller blocks (but still impressive) in building their Temple of Jupiter on the ancient spot, but notice that the Roman blocks are filled with "Lewis holes" used to lift heavy stone blocks. The megaliths below are free of such holes.

In 1977, Jean-Pierre Adam went into detail about how the blocks might have been moved from the quarry, where another 1200 ton rock, the "Stone of the Pregnant Woman," remains unused.[12] He points out that St. Catherine II of Russia had a massive block brought to St. Petersburg to serve as the base of the "colossal" equestrian statue of Peter the Great. Adam says, (my translation from French): "This is most likely the largest stone ever moved by man, one and a

---

12 Adam, J (1977). A Propos du Trilithon de Baalbek. Le Transport et la Mise en Oeuvre des Megaliths. *Syria*, 54(1-2): 31-63.

A Worldwide Flood?

Figure 26: Temple of Jupiter in Baalbek, Lebanon, from Bonfils, successor A. Guiragossian, Beirut Collection of the Vues de tout l'Orient. The two men standing in the crack in the wall provide scale.

Figure 27: Stone of the Pregnant Woman, a giant stone that remains cut, but unmoved, from an ancient quarry in Baalbek, Lebanon. The man on the stone provides scale. Photo by Ralph Ellis.

half times the weight of the trilithon blocks."[13]

The bottom line is this: whether we know *how* something was done has nothing to do with anything. The Baalbek stones were set into place long ago, the work of clever people with serious skills.

In other words, we can't say a thing about Noah's ark with so little knowledge of our own. We don't even know what "gopher" wood was, or if it was even *wood*. We can imagine answers to questions about ventilation, refuse removal and so forth, but we don't even know what the insides of the ark looked like. We don't know how the animals behaved or Noah's skill with them. If the Bible's narrative is accurate about the early Earth, the biological community was a completely different organism back then. We're sitting around, arguing over things we don't know a thing about.

At the same time, the Bible's narrative gives an internally consistent description of what would happen if a pristine world became corrupted and broken. There are a variety of details that support this ongoing theme. It sketches an interesting picture for us, and I want to take a brief look.

THE AGES

One thing that interests me are the human ages. Before the Flood, Noah's forefathers almost all lived more than 900 years. According to Genesis 3:22, Adam had the potential to live forever. He and Eve got kicked out of the Garden and couldn't eat the Tree of Life, but they still lived past the age of 900. That seems ridiculous to us today, but it's consistent with what we'd expect to see if Genesis were actual history. If the early humans were genetically perfect, their bodies would have operated with perfect efficiency. Their sons could have married their daughters without any fear of compounding genetic defects. Whatever the firmament was, it appears to have protected the planet and the creatures in it, because the environment changed drastically after the Flood. Noah continued to the age of 950,[14] but his son Shem died at just 602,[15] and the ages of Shem's

---

13 Ibid., 42.
14 Genesis 9:29
15 Genesis 11:10-11

descendants listed in Genesis 11 drop dramatically after the Flood. Abraham survived to 175 and Moses died at 120, and yet even in Moses' time, 80 was considered old.[16]

We see the same pattern in the Sumerian King List. The ages are even more ridiculous, but they follow the same *pattern*. The first king Alulim ruled for 28,800 years, followed by Alaljar who ruled for 36,000 years. After the Flood, though, the ages tank. Jucur ruled for 1,200 years and Kullassina-bel ruled for 960 years. It doesn't take long for the ages to fall into what we'd consider a reasonable range.

VEGETARIANISM

One of the big questions skeptics ask is, "How did Noah keep the meat eaters from eating the other animals?" According to Genesis 1:29-30, the animals were all vegetarians at the beginning, and animals did not fear humans until Genesis 9:2 - after the Flood. (By the way, Isaiah prophesies that these conditions will return one day.[17]) Rooms would be valuable in an ark full of animals, but cages wouldn't be necessary. After the Flood, God gave Noah permission to eat the animals,[18] which implies that Noah was also a vegetarian before the Flood. If the destruction of the firmament resulted in a quickly deteriorating world, then animal flesh would have provided an alternate source of nutrition.

THE RAINBOW

Then, there's the rainbow. It's easy to say, "Oh, this is just the Hebrew explanation for the existence of rainbows," except that it doesn't make any attempt to explain where rainbows come from. Genesis offers no stories about Stork painting the sky or demigods throwing their hair from celestial chariots. There's no imaginative tale involved here. Instead, the Bible says there was no rain before the Flood. Then, after the Flood, a rainbow appeared in the sky.[19] That's just descriptive. I wouldn't think Noah understood about the

---

16 Genesis 25:7; Deuteronomy 34:7; Psalm 90:10
17 Isaiah 11:1-10; 65:17ff
18 Genesis 9:3
19 Genesis 2:5-6; Genesis 9:13-17

refraction of light through water droplets in the atmosphere, but the existence of that rainbow as a new thing tells us that something significant had changed meteorologically.

GENES AND SIBLINGS

There are other things. Abraham - one of the very greatest men in the Bible - married his half-sister. People married their close relatives in the ancient world. By the time of Moses, it was determined that marrying siblings was a bad thing.[20] The whole reason we shouldn't marry our cousins is the risk of compounding family genetic defects. If Adam and Eve were created with pristine DNA, Cain and Seth could have produced perfectly healthy children with their sisters. As one generation followed another, though, errors would have crept into the gene pool, and it would have become increasingly important to avoid having offspring with close relatives who shared the same destructive mutations.

What we see in the Genesis narrative is what we'd expect to see if the first humans were created in perfection and genetic errors crept in over time, especially after the Flood drastically changed the environment.

So. Which is it? Did we start as amoebas and develop into all the wide variety of life on earth - or - has natural selection simply worked as a protective mechanism to stave off complete extinction as we slip and slide from our once-glory?

The fact is, the Baalbek stones and the pyramids and other ancient mysteries support the idea that our ancestors were brilliant and capable. Geometry and math and astronomy were highly developed in ancient Sumer and Babylon, long before us. Here in the 21st century, we've developed fantastically advanced technology, but we're also all dying of cancer and diabetes and heart disease. If humankind had clean, sharp minds and strong bodies and people lived many hundreds of years, there's no reason to think those early humans couldn't have schooled us and slandered our mamas.

---

20 Genesis 20:12; Leviticus 20:17

A Worldwide Flood?

At the time of this writing, the Human Genome Research Institute at the National Institutes of Health is busily populating a database of human diseases caused by damaging mutations in the genetic code. There are plenty of neutral mutations too, but a multitude of dangerous or painful or inconvenient disorders are caused because somebody's DNA has an error.

As far as I know, there are no databases collecting all the wonderful new superpowers that people have received because their genetic codes got changed up - no databases describing the wings, the new organs, the fire breathing capability, the additional eyes or radar hearing of mutants. The X-Men aren't real.

These facts don't answer our questions about geology or ark survival, but it's unwise to dismiss the consistent picture that Genesis gives us.

On the other hand, I can ask a different question: "How could bacteria hibernate in a mastodon's guts for 11,000 years, let alone in halite for 250 million years? Wouldn't the DNA and proteins break down?"

And that right there is the bottom line problem with these sorts of controversies. We have to be cautious with our feasibility studies when we cannot recreate the ridiculous magnitudes required. We cannot observe bacteria for 250 million years. We cannot recreate the conditions of a global flood. It's not just the assumptions we have to make; we don't have any way to recognize when certain important factors have been overlooked.

Does that mean we give up? Of course not! It just means we proceed with a smidgen of humility. That... that's what I think is my biggest point. It's what I would hammer on repeatedly to those on all sides of an issue. Whatever *is* the true history of the Earth, we're all in the same boat, folks.

WHO WROTE GENESIS?

This is an important question. What is Genesis? Is it the Word of God spoken to Moses? Is it a collection of Jewish myths that represent the evolution of the Jewish religion, or is it a true record of the birth of humanity?

The first five books of the Bible are called the Law, or the Torah, or the Pentateuch. Five books. Genesis, Exodus, Leviticus, Numbers, and Deuteronomy. They are known as the books of Moses, because throughout history, everybody recognized they were written by Moses, the servant of God.

If that's not true, then the whole thing is a human invention, and we can take it for whatever it's worth. If it *is* true, then we need to take Genesis seriously, because it's the very Word of God. It's an important controversy, and the correct answer *matters*.

Let's start with Moses' special relationship with God.

In Numbers 12, we read a warm side story about Miriam and Aaron, the sister and brother of Moses. It starts because Moses has married an Ethiopian woman, and Miriam and Aaron think it's their job to criticize Moses about it. They insist they've got the right to confront Moses, spouting off in Numbers 12:2, *"Has the LORD indeed spoken only through Moses? Has He not spoken through us also?"*

Yeah, a little attitude. I'm pretty sure Miriam is the main antagonist here. That sounds right, based on what we know about Aaron and Miriam. Miriam is the sort to take it on herself to follow her baby brother Moses down the river. Aaron...Aaron gives in to pressure. Remember when he built the Israelites a golden calf in Exodus 32 while Moses was up on Mount Sinai? It's clear he and Miriam have two different personality types.

When Miriam and Aaron start berating Moses, the LORD quickly humbles them both. God says, "All three of you, come here," as though they're children having an argument. He orders them to the

Tabernacle, where He puts a halt to Miriam and Aaron's arrogance:

And the LORD spake suddenly unto Moses, and unto Aaron, and unto Miriam, Come out ye three unto the tabernacle of the congregation. And they three came out. Then the LORD came down in the pillar of cloud and stood in the door of the tabernacle, and called Aaron and Miriam. And they both went forward. Then He said, "Hear now My words: If there is a prophet among you, I, the LORD, make Myself known to him in a vision; I speak to him in a dream. Not so with My servant Moses; He is faithful in all My house. I speak with him face to face, Even plainly, and not in dark sayings; And he sees the form of the LORD. Why then were you not afraid To speak against My servant Moses?"

Numbers 12:4-8

If God has a problem with the fact that Moses married an Ethiopian woman, God can say so Himself. Moses is not spiritually hard of hearing. God doesn't need Miriam and Aaron's help, so why did they feel free to butt in? Bad move. When the cloud leaves, Miriam's skin has become leprous. In a panic, Aaron apologizes, and Moses asks God to heal her. God responds.

Then the LORD said to Moses, "If her father had but spit in her face, would she not be shamed seven days? Let her be shut out of the camp seven days, and afterward she may be received again."

Numbers 12:14

I read that as a child and thought, "Lord, I never ever want You to feel like spitting in my face." Miriam was still His daughter, His servant, but He was disgusted with her. You can't have attitude around God. Human arrogance is a stern no-no.

Which brings us to the question of who wrote this story in the book of Numbers in the first place.

The Torah and Talmud

The text itself says that Moses wrote it all. The Torah straight up tells us that Moses wrote down the Law and the history of the Exodus and all the wanderings in the desert. I've added some bold to the following verses to aid the skimmers among us:

So Moses came and told the people all the words of the LORD and all the judgments. And all the people answered with one voice and said, "All the words which the LORD has said we will do." **And Moses wrote all the words of the LORD...**

Exodus 24:3-4a

Then **he took the Book of the Covenant** *and read in the hearing of the people. And they said, "All that the LORD has said we will do, and be obedient."*

Exodus 24:7

Now **Moses wrote down the starting points of their journeys** *at the command of the LORD. And these are their journeys according to their starting points:*

Numbers 33:2

So **Moses wrote this law** *and delivered it to the priests, the sons of Levi, who bore the ark of the covenant of the LORD, and to all the elders of Israel.*

Deuteronomy 31:9

So it was, **when Moses had completed writing the words of this law in a book,** *when they were finished, that Moses commanded the Levites, who bore the ark of the covenant of the LORD, saying: "Take this Book of the Law, and put it beside the ark of the covenant of the LORD your God, that it may be there as a witness against you;*

Deuteronomy 31:24-26

According to the Torah itself, Moses wrote down the legal codes that God gave him in Exodus and Leviticus, the wilderness wanderings in Numbers, and a repeat of the law of Deuteronomy, and the whole thing is called the Book of The Law. There are certain phrases that tell us somebody came behind and put it into its final form; for instance, Moses likely didn't write about his own death:

> *So Moses the servant of the LORD died there in the land of Moab, according to the word of the LORD. And He buried him in a valley in the land of Moab, opposite Beth Peor; but no one knows his grave to this day.*
>
> Deuteronomy 34:5-6

Clearly Moses didn't write that part. What's more, the statement, "...no one knows his grave to this day," tells us it was written some years after the fact. The summary of the death of Moses finishes up the Torah, and the book of Deuteronomy ends a few verses later.

Traditionally, it's understood that Joshua had the honor of finishing out Deuteronomy. Exodus 24:13 tells us Joshua went up on Mount Sinai with Moses when he went to get the Law from God. In Deuteronomy 31:7,14, Moses inaugurated Joshua as the next leader, and we see in Joshua 24:26 that Joshua felt free to continue the story, to add to the Book of the Law the events that happened in his day.

Aside from the Bible itself, the Talmud agrees that Moses wrote the Torah and Joshua wrote both the Book of Joshua and the last eight verses of Deuteronomy:

> ...Moses wrote his own book, i.e., the Torah, and the portion of Balaam in the Torah, and the book of Job. Joshua wrote his own book and eight verses in the Torah, which describe the death of Moses...
>
> Bava Batra 14b [21]

Okay. Got it.

This is the view of the ancient rabbis. It was the consistent view,

---

21 Translation from The William Davidson digital edition of the *Koren Noé Talmud*.

one that the rabbis claimed was passed down from generation to generation from the time of the Exodus.

Frankly, if anybody in the world was capable of writing an accurate chapter on how the world was made, it would have been Moses, who talked to God face to face.

INTERNAL EVIDENCE

In the next few pages, we'll hit the Documentary Hypothesis - the modern view that lots of different people (not Moses) wrote the Torah. Before we even get there, I want to point out something super interesting, something that supports the case that a single person wrote all five books. I'm doing it here, right up front, so in case you get distracted you'll remember this fact at least. The Pentateuch text never uses the word "she." It uses the word "he" for both men and women. Herbert Wolf kindly points this out for us:

> Along with the overall continuity in the narrative we can also point to the grammatical features that underscore the unity of the Pentateuch. For some reason these five books fail to distinguish between the third person pronouns, "he" and "she." Instead of using *hu* and *hi* like the rest of the Old Testament, the Pentateuch uses only the masculine form.[22]

Moses, that male chauvinist! Just kidding. Languages like Finnish and Bengali don't distinguish between male and female pronouns either, and it's a notable grammatical characteristic of the Pentateuch that separates it from the rest of the Old Testament writings. If lots of people wrote the Pentateuch, we might expect the feminine pronoun to show up now and then sporadically. If it were put together late in Jewish history, we'd expect the editor to err on the side of fixing the pronouns, because the rest of the Hebrew Scriptures use both "she" and "he." No. The Torah consistently uses only male pronouns, and that points to a single author, (and one nobody dared to correct).

There's another reason I think Moses wrote the Pentateuch: I keep a journal. While I can cover the history of my ancestors in

---
22 Wolf, H. (1991). *An Introduction to the Old Testament Pentateuch* (p. 20). Chicago: Moody.

a chapter or two, my journals fill volumes. Why? Because I have to research my ancestors' histories, but I'm present to write about my own life. That's what we see in the Torah. Genesis covers some 2250 years of human history - in one book - but the 40 years of the Exodus and wilderness wanderings took four additional books to describe. Why so much time spent on Exodus, Leviticus, Numbers and Deuteronomy? Because Moses was there to write the words of God and the events taking place, and he had a lot to say. That's what happens when a person who is *there* starts writing it all down. And note: as soon as Moses died, his narrative ended.

So, what's the issue? Why is there any question about the Mosaic authorship of the Pentateuch?[23]

IBN EZRA AND SPINOZA

For millennia, Jews and Gentiles alike took it for granted that Moses had written the first five books of the Bible. The renowned 12th century Rabbi Abraham Ibn Ezra did wonder how Moses wrote about his own death in Deuteronomy 34:5-12. He also wondered why Moses said, "And the Canaanites were then in the land," in Genesis 12:6 when Abraham showed up there. The 17th century Dutch-Jewish philosopher Baruch Spinoza interpreted this passage as though it were written long after Joshua fought off the Canaanites, and he used this phrase to argue that Moses hadn't written the Torah after all. Spinoza didn't believe in miracles or angels or God's supernatural intervention in the world, and he blatantly argued that the Pentateuch had been written long after Moses died. The Jewish leadership in Amsterdam eventually excommunicated Spinoza as a pantheist.

I think Spinoza read Genesis 12:6 wrong. I understand it to mean that the Canaanites had moved *into* the future Holy Land by that point. Canaan was the grandson of Noah, on whom Noah put a curse in Genesis 9:25. (He didn't put a curse on Mizraim or Phut or Cush, Ham's other children. Just Canaan.) According to

---

23 I once took a class dedicated to the Pentateuch, reading passages in Hebrew and taking into account the histories of the Egyptians and Hittites and Assyrians. Despite these serious things, the word "Pentateuch" makes me want to call out "Peregrin Took!" in a very angry Gandalf voice. Forgive me.

Genesis 12:4-6, Abraham headed into the land of Canaan when he was 75-years-old, which was fewer than 400 years after the Flood. By that time, the Canaanites had already settled down there. This alternate interpretation makes more sense, because the Canaanites were never completely kicked out of the land after Joshua. In fact, Zechariah 14:21 indicates they were still an issue at the end of the Old Testament.

So, that's a lot of wind and noise over a non-issue.

JEAN ASTRUC

The French researcher Jean Astruc agreed that Moses wrote the Pentateuch, but he suggested that Moses had taken advantage of earlier manuscripts in compiling Genesis. In 1753, he published a book with the exceptionally long title: *Conjectures about the Original Memoirs which it Appeared that Moses Used in Composing the Book of Genesis with Certain Remarks which Help Clarify these Conjectures.*

Astruc suggested that Moses accessed earlier writings to produce the history of the Hebrew people prior to the Exodus, and there's a case to be made for this idea. Even if Moses talked face to face with God, that doesn't mean he didn't have access to sources that described the history of the Hebrew people. Eleven times in Genesis we find the phrase: "These are the generations of..."[24] from the Hebrew word *toledoth*, which has to do with offspring and descendants. Connotatively, though, the word toledoth can mean "history." In Genesis, the word is almost always followed by a genealogy, but each of these toledoth statements also begins (or ends) a new section of Genesis. They provide a series of accounts about the events that took place from the Creation until Jacob's 12 sons moved their families into Egypt. It may be that the word toledoth was Moses' way of tying different historical documents together. We also find statements like, "This is the book of the generations of Adam," in Genesis 5:1, which suggests that Moses accessed a *document*. If Moses used earlier accounts to compile Genesis, that didn't prevent God from weighing in.

---
24 Genesis 2:4; 5:1; 6:9; 10:1; 11:10; 11:27; 25:12; 25:19; 36:1; 36:9; 37:2.

Astruc specifically saw differences in the first two chapters of Genesis. He noticed that the name *Elohim* was used for God in Genesis 1, but *Yahweh Elohim* (commonly translated as "the LORD God") was the name used in Genesis 2. Astruc thought this indicated that Genesis 1 and 2 were two different creation accounts that Moses had put together.

There are unique differences between Genesis 1 and 2, and we read about locations named after the direct descendants of Noah as named in Genesis 10. The writer was referring to places his readers would know about - Cush, Havilah, and Asshur - which existed during the time Moses would have written.

Genesis 1 and 2

I agree these chapters offer us two different accounts, but I read Genesis 1:1 – 2:3 as the account God gave Moses directly and Genesis 2:4-4:26 as the record from Adam's point of view, as redacted by Moses. God and Moses had face-to-face discussions for 40 years, and Moses produced Chapter 1 as a description of God's great power in bringing order from chaos. Chapter 2 is more personal. We see the Lord forming Adam from dust, hands-on. Adam may have developed a written language during his 930 years, but at the least he could spoken to multiple generations of descendants, even to Noah's grandpa Methuselah and dad Lamech.

Genesis 1 gives the name for God as *Elohim*. It's the plural form of *El*, the generic word for "god" or "powerful one." Elohim is the all-powerful, ultimate God of the universe. This name has the connotation of somebody who wields great authority and might. Interestingly, though, when the Bible calls God "Elohim," it uses singular verbs, so that a plural word is treated as a single person.

Starting in Genesis 2:4, the name of God is given as *Yahweh Elohim*. To Elohim has been added the personal name of the Hebrew God, the name given to Moses at the burning bush in Exodus 3.

We don't know exactly how to pronounce the original name Yahweh, because the Jews were afraid to say it out loud. They replaced it with "the LORD" out of respect, and the rest of us have followed suit.

Right. So we have "God" in Genesis 1 and "Yahweh God" in Genesis 2. As we read through the Pentateuch, we sometimes find "God" favored and sometimes "Yahweh" favored, and sometimes they're mixed together in one verse, like in Exodus 3:4 or 3:15:

> *And* **God** *said moreover unto Moses, Thus shalt thou say unto the children of Israel, The* **LORD God** *of your fathers, the God of Abraham, the God of Isaac, and the God of Jacob, hath sent me unto you: this is my name for ever, and this is my memorial unto all generations.*
>
> Exodus 3:15

In one verse, we see God referred to as both "Elohim" and "Yahweh Elohim," which tells us the writer was willing to interchange these names for God without issue.

Yahweh is a great name for God. I love it. It means, "I AM." As opposed to all the other gods of the nations, Yahweh is the God who actually exists. What's more, it's in a form that indicates that He exists continually. On and on and on. In John 8:58, Jesus makes the famous statement, *"Before Abraham was, I am,"* which enrages His enemies. They pick up stones to stone Him, because they know exactly what He's saying. I bring this up for two reasons: first, Jesus claimed to be the God of the burning bush, and second, He used the term "I Am" as an on-going existence that extended from the deep past up to His present day. Cool stuff.

BIBLE NAMES

There's been a tendency among skeptics to regard different names of God in the Bible as evidence of different authors who called God by their own favorite names - or who worshipped different gods altogether. It's important to understand how names are used in the Bible. In our culture, names are primarily identification labels. Names meant *more* in the Bible; they said something about the person himself. It's the difference between calling a guy, "Hammer" as a cool name tag and calling him, "Hammer" because he goes around knocking people out. "Hammer" tells you something about the guy who was given that nickname. Names matter. God changed Abram's name to Abraham, "father of a multitude," because vast nations were going to come from him. God changed Jacob's name to "Israel," because he struggled with God and men and prevailed.

God's names in the Bible often have a lot to do with what He's *doing*. We see this in Isaiah. Isaiah calls God by a variety of titles, but they always say something about the role God is taking in the passage. If He's directing the armies of Israel, Isaiah calls Him, "Yahweh of Hosts." If Isaiah is portraying God's tender care of Israel, he calls Him, "Lord Yahweh."[25] If Isaiah is focusing on God's mighty power and righteousness, he gives Him the distinct name, "the Holy One of Israel."

It's a mistake to assume that different names of God indicate different authors. The root of Elohim is *el*, which is all about power. In Genesis 1, God is demonstrating His great power in bringing order out of chaos and creating the universe from nothing. In Genesis 2-3, however, we find God taking a personal role with His human creations. He is Yahweh God, down in the dirt, talking one-on-one with the beings He's made.

Moses also gave multiple names to some humans, which is used to reject his authorship of the Torah. I think that's *so* silly. Most of the time, the Torah refers to Moses' father-in-law as "Jethro," but in Exodus 2:18 and Numbers 10:29, he's called "Reuel." Multitudes of "Jethro" and two instances of "Reuel." Jethro ("Excellence") may have been his title and Reuel his given name, but he also might have had two names used interchangeably, like Jacob/Israel. There's also the complaint that the Torah mostly calls the Canaanites "Canaanites" but also calls them "Amorites." The Amorites were one group of Canaanites, so both names were appropriate. Joshua refers to both Amorites and Canaanites within two verses in Joshua 7:7-9.

---

25   The Hebrew that is literally translated as "Yahweh of hosts" is generally given as "LORD of hosts" in most Bible translations. Similarly, "Lord Yahweh" is often translated as "Lord GOD" or "Sovereign LORD."

Many people are called by different names. The lead singer for U2 is Bono. Everybody knows him as Bono, so he knows that if he gets a call for "Paul David Hewson" it's just a telemarketer with no clue. He was asked once what his wife calls him, and he answered, "She calls me 'Baby.'"

Human writers are multi-faceted. They're not androids with simple programming. Moses had to spit desert dust out of his teeth every day, but that didn't make him one-dimensional. He can call his father-in-law "Reuel" if he wants.

JEDP: The Documentary Hypothesis

Of course, certain textual critics come to the Bible with the view that human beings don't live 930 years, isochron dating indicates that the world is better than 4.5 billion-years-old, and God doesn't intervene in human lives – if He exists at all.

It became more and more acceptable in the 19th century for intellectuals to approach the Bible with the attitude that it was simply a religious book – a compilation of Jewish mythologies faithfully preserved for generations. A variety of dissenting voices suggested that the Book of Genesis was a mishmash of materials compiled long after the events of the Exodus (if a literal Exodus took place at all).

One of these voices was the German scholar Julius Wellhausen (1844-1918). Wellhausen made famous the Documentary Hypothesis, the idea that the Pentateuch was a collection of writings by different groups of people over hundreds of years. Karl Heinrich Graf (1815-1869) had argued that the J and E sources came first and the P source came last, and so this view is often called the Graf-Wellhausen Documentary Hypothesis.

Wellhausen and others agreed that the use of "LORD" and "God," stemmed from different writers of the original texts. He saw creation accounts, histories, genealogies, legal documents, poetry – all different types of writing – and argued that the Pentateuch was not written by Moses under the inspiration of God but was instead compiled from at least four different sources over the course of centuries. These were designated as J (Jahwist), E (Elohist), D

(Deuteronomist), and P (Priestly) texts. Wellhausen said that J was added to E, and then JE to D, and then P was finally added by redactors during the time of Ezra about 450 BC.

- **The J Source:** These were the sections of the Bible (sometimes as small as a partial verse) that were written by those authors who called God by the Hebrew name Jahweh (Yahweh). The J source documents were given a date of 900–850 BC, the century after the death of King Solomon.
- **The E Source:** These were the sections of the Pentateuch that used the name Elohim and were supposedly written around 750–700 BC.
- **The D Source:** This was a legal document allegedly written during the time of Josiah's reforms in the late 7th century BC.
- **The P Source:** These came together as the legal codes, especially in Leviticus, and were likely written by priests after the Temple was destroyed and the Jews were exiled to Babylon in 586 BC.

According to the Documentary Hypothesis, these four sources were once continuous documents of their own, and later redactors combined them together, which resulted in repetitions and varied vocabulary that reflected the multiple sources. The Documentary Hypothesis grew popular in the late 19th century, and source critics tediously worked through the Torah and divided it up between these imagined sources. They chopped up continuous passages, sometimes splitting a verse and assigning half verses to different sources. The JEDP sources were further subdivided, until it became hard to keep track of the cooks in the literary kitchen. Carpenter and Harford-Battersby list a colorful, flavorful array of divisions within the J E D and P schools, along with various redactors.[26]

---
26 Carpenter, J.E. and Harford-Battersby, G. (1900). *The Hexateuch According to the Revised Version* (p. xii). London: Longmans, Green, and Co.

The first problem with this multiple-source approach is that it's massively subjective, based on a philosophical view that the Israelite culture evolved from nothing-to-something slowly over centuries after the Davidic monarchy had already developed. This view has no direct evidence from any other documents – no actual P source or J source. What's more, it blatantly denies the Israelites' documented history. Israelite documents declare, "We serve the God of Abraham, Isaac, and Jacob who rescued us from slavery in Egypt. And Moses wrote all of this down." The Pentateuch explains exactly how the Israelite people and religion began. Substantial evidence, not mere speculation, should be necessary to reject this documented history.

ANCIENT INTERNATIONAL RELATIONS

It used to be argued that Moses couldn't have written the Pentateuch because there was no writing in those days. It's been assumed that if he did write, he wrote in Egyptian. It turns out there was plenty of writing before Moses' day.

Hittite treaties dating as early as 1500 BC were discovered in the royal archives of Boghazkoi in Turkey. A law historian named Victor Korošec (1899-1985) analyzed Hittite treaties of the 2nd millennium BC,[27] and Bible scholar George E. Mendenhall (1916-2016) compared those suzerainty treaties with the covenants of the Pentateuch and found similarities.[28] Deuteronomy has a structure similar to a 2nd millennium BC vassal-lord treaty, complete with curses and blessings.

That's great, but it's not as interesting to me as the very fact that Hittites were writing sophisticated legal documents in the 2nd millennium before Christ. (The Sumerians had been doing it for centuries already, because wars and bureaucrats have been around as long as civilizations.) These sorts of treaties were being used between nation states all over the Middle East. The Amarna Tablets give us tremendous insight into the diplomacy between the Egyptians and Canaanites/Amorites. The Amarna Letters are primarily written in

---

27 Korošec, V. (1931). *Hethitische Staatsverträge : ein Beitrag zu ihrer juristischen Wertung*. Leipzig: Weicher.
28 Mendenhall, G. (1955). *Law and Covenant in Israel and the Ancient Near East*. Pittsburgh: Biblical Colloquium.

Akkadian cuneiform, not Egyptian.

One of my favorites is the treaty between the Egyptians and Hittites after the famous Battle of Kadesh near today's Syria-Lebanon border. The Egyptians and Hittites fought over who would control Amurru (the Amorites), and both sides claimed victory in their respective accounts. Archaeologists have recovered two copies of the treaty, the first in Egyptian hieroglyphics and the other in Hittite-Akkadian. That's fantastic from an archaeological perspective, because it allows a comparison of accounts between the two people groups.

For our purposes here, these kinds of treaties show that there were a variety of written languages running around during those years. The people of Cyprus had a chicken scratch of their own. Proto-Sinaitic / Proto-Canaanite was an alphabet-based form of writing that developed in the lands of the Bible somewhere between 1850 BC and 1550 BC and is regarded as an ancestor to the Phoenician alphabet.

Here's my point. If Moses spent 40 years in Pharaoh's court, he was not only educated and able to write, but he would have known several languages. He could have communed with people in the surrounding nations and written in the various forms of script.

Culture Shock

So Julius Wellhausen and his compadres in the West decided they were qualified to imagine groups of authors, then dissect the Torah between these imaginary groups based on those groups' assumed purposes for writing. The 19th and 20th century European scholars like Wellhausen had a real problem in doing this: they presumed to determine the purposes of ancient Hebrew works written in an entirely different culture on a different continent in a different millennium. They ignored the text's statements that Moses wrote it all down, the use of male pronouns throughout, the unanimous agreement in the ancient world that Moses was the author. They did those things on purpose, because they thought they knew better.

That's problematic, certainly, but what's really embarrassing is how careless the Documentary Hypothesis people are with the artistic style of Hebrew literature. It's like they don't understand certain basics.

For instance, textual critics have pointed to repetitions in Genesis as evidence for multiple sources spliced together. Their idea is that two different texts about the Flood were pieced together by redactors, and some of the same elements were repeated when this was done.

Unfortunately, it's a super foolish view to hold about repetitions in Hebrew literature. Hebrew authors often repeated ideas for emphasis. Repetitions and doublets and parallel accounts are common rhetorical devices in ancient Hebrew texts, and they point to a *single* author because they were carefully, purposefully developed. Every Christmas we sing a text from Isaiah 9:6:

Unto us a child is born. Unto us a son is given.

Thanks to Handel for the excellent music, but Isaiah purposely created that couplet. Isaiah loved couplets as a rhetorical device.

Isaiah 53:5 states, "*But He was wounded for our transgressions, He was bruised for our iniquities…*"

The textual critics who chopped up Genesis into J and E and P sources didn't seem to understand this practice. According to the Documentary Hypothesis, Genesis 6:9–22 belongs to the (newer) P source, while 7:1–5 has been assigned to J, with the assumption that a single author wouldn't have had God repeat Himself by telling Noah two different times to bring pairs of animals into the ark. The conclusion: two different Flood stories were smooshed together.

That's shortsighted. The LORD gave Noah a general set of instructions in chapter 6 to get him to build the ark, and 100 years later He said, "Okay. Let's load everybody up. Let's do this thing." God's repetition of what He said earlier heightens the reader's anticipation of entering the ark as chapter 7 begins. It's purposeful.

The Hebrew scriptures often use a literary structure called a chiasm, a mirroring of ideas in reverse order. We're familiar with palindromes like MOM or TACOCAT or STEP ON NO PETS, and Chiastic structure follows that same kind of repeated form, the same forward and backward. Its mirror-image structure is found in highly stylized, carefully constructed narratives in the Bible.

Again, Genesis 7:21 has been assigned to P and Genesis 7:22 to J, because there's an echo in content:

And all flesh died that moved on the earth: birds and cattle and beasts and every creeping thing that creeps on the earth, and every man.

Genesis 7:21

All in whose nostrils was the breath of the spirit of life, all that was on the dry land, died.

Genesis 7:22

This is clearly a chiastic couplet used to describe a great, very big, huge calamity, and it points back to a similar statement by God in Genesis 6:17, bringing the story full circle.

Genesis 7:21-22
A. All flesh that moved on the earth died,
   B. birds and cattle and beasts and creeping things,
      C. and every man,
   B'. all in whose nostrils was the breath of life,
A'. all that was on the dry land died.

This is common in Genesis, and Bible scholar Francis Andersen explains, "The rhetorical effect of this kind of epic repetition is to slow down the pace of the narrative. It holds the picture a little longer and enforces it on the mind."[29] The Flood narrative was written in a highly artistic fashion that points solidly to a single writer - not at all sloppy cutting and pasting from multiple sources.

---
29 Andersen, F. (1974). *The Sentence in Biblical Hebrew* (p. 40). The Hague: Mouton Publishers.

The Pentateuch Outside The Pentateuch

In the Bible, we find that the Torah is mentioned early on, from the time of Joshua into the Monarchy and throughout the Prophets. "The Law" always referred to those first five books. It was given authority right away, and Moses was the only one credited with writing it:

> *Only be strong and very courageous, that you may observe to do according to all the law which Moses My servant commanded you; do not turn from it to the right hand or to the left, that you may prosper wherever you go. This Book of the Law shall not depart from your mouth, but you shall meditate in it day and night, that you may observe to do according to all that is written in it. For then you will make your way prosperous, and then you will have good success.*
>
> Joshua 1:7-8

There's no confusion. There's no deviation. Moses is the assumed writer of the Torah throughout the New Testament as well, and Jesus consistently confirms that Moses is the writer:

> *And Jesus said to him, "See that you tell no one; but go your way, show yourself to the priest, and offer the gift that Moses commanded, as a testimony to them."*
>
> Matthew 8:4 (Cf. Leviticus 14:1–32)

> *For Moses said, 'Honor your father and your mother' and, 'He who curses father or mother, let him be put to death.':*
>
> Mark 7:10 (Cf. Exodus 20:12, Leviticus 20:9)

> *And as touching the dead, that they rise: have ye not read in the book of Moses, how in the bush God spake unto him, saying, I am the God of Abraham, and the God of Isaac, and the God of Jacob?*
>
> Mark 12:26 (Cf. Exodus 3:6)

> Then He said to them, "These are the words which I spoke to you while I was still with you, that all things must be fulfilled which were written in the Law of Moses and the Prophets and the Psalms concerning Me."
>
> Luke 24:44

It matters whether Moses spoke to God face to face. If he didn't, then his words are merely ancient Israelite ideas, which we are free to accept or reject. If he did, though, then they are the very words of God, and we need to pay attention.

If Jesus isn't God, then He was only quoting a law written by His ancestors. But, if He is God, if He is the I Am made flesh, the incarnate LORD of the burning bush, then these things absolutely matter, and we'd better get it right.

> "For if you believed Moses, you would believe Me; for he wrote about Me. But if you do not believe his writings, how will you believe My words?"
>
> John 5:46-47

WHAT IS THE ARCHAEOPTERYX?

Joe David and I watched *The Lost World: Jurassic Park* in the theaters in 1997, and I felt warm and gushy because I had a piece of *Triceratops* bone tucked into my pocket. We didn't care that most of the dinosaurs in the film were from the Upper Cretaceous; we had fun watching mommy T-Rexes chase Jeff Goldblum and Vince Vaughn. Of course, Stephen Spielberg's crew continued to create *Velociraptors* that were too big. The *Velociraptors* in the movie are based on a larger relative, *Deinonychus* ("terrible claw"), but author Michael Crichton hadn't liked the name *Deinonychus*, so he conflated it with the smaller, golden retriever-sized *Velociraptor*. More than that, the movies left these brutal killers all smooth and naked! So embarrassing for everybody.

This is important. While the *Velociraptors* in the Jurassic movies had bare, scaly skin, the presence of "quill knobs" on the bones of their forearms suggests they wore feathers in real life.[30] At the very least, a whole array of little, two-legged dromaeosaurid skeletal remains have been found surrounded by feather markings, and *Velociraptor* and *Deinonychus* are genera in the happy Dromaeosauridae family. They have a home. They belong.

Figure 28: Cast of left hind foot of Deinonychus antirrhopus *(holotype YPM 5205) (Ostrom 1969). Adapted from photo by Didier Descouens. Reproduced under international license CC BY-SA 4.0.*

Some dromaeosaurids had long arms and some had short arms, some had more feathers and some had fewer. They all had that distinctive sharply-curved, hyperextensible "sickle claw" on each foot that could have been used to eviscerate its prey. Remember when the children were hiding in the kitchen in *Jurassic Park*, and the *Velociraptor* tapped that sickle-claw on the kitchen floor? Run away! Run

---

30 Turner, A., Makovicky, P., &Norell, M. (2007).Feather Quill Knobs in the Dinosaur Velociraptor. *Science,* 317(5845), 1721-1721.

What is the Archaeopteryx?

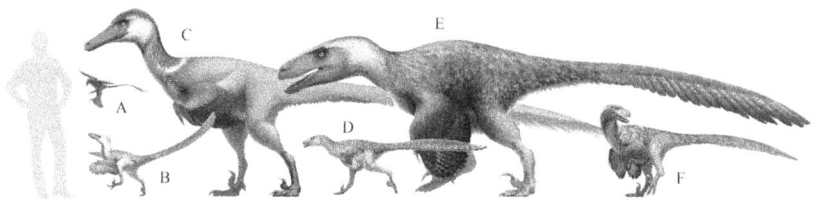

Figure 29: Size comparisons of well-known dromaeosaurs: (A) Microraptor gui, *(B)* Velociraptor mongoliensis, *(C)* Austroraptor cabazai, *(D)* Dromaeosaurus albertensis,*(E)* Utahraptor ostrommaysorum, *and (F)* Deinonychus antirrhopus. *Adapted from Fred Wierum. Reproduced under international license CC BY-SA 4.0.*

away! Science artists now paint *Velociraptor* as a vicious bird-like creature, one that paleontologist Stephen Brusatte has called, "a fluffy, feathered poodle from hell."[31]

In 2015, Brusatte and his associate Junchang Lü published on *Zhenyuanlong suni*, a big, winged dromaeosaurid from the Lower Cretaceous.[32] This particular specimen is interesting because it was larger with small, feathered wings. Like an ostrich, it wouldn't have been able to fly, but it wore large, quill-like pennaceous feathers.

It's not just the dromaeosaurids. Tyrannosaurs may have had some feathering going on. The basal tyrannosauroid *Yutyrannis huali* was covered by fuzzy long filamentous feathers.[33] A juvenile megalosauroid in Germany was found with filamentous feathers on parts of its body.[34] A variety of feathered dinosaurs have been found in China,[35] and it seems that feathered dinosaurs were fairly common. In fact, even an ornithischian dinosaur called *Kulindadromeus zabaikalicus* was discovered in Siberia with evidence

---

31 Brusatte, S. (2015, July 16). "Discovering a New Dinosaur Helped Us Prove Velociraptors Had Feathers." Retrieved from https://theconversation.com/discovering-a-new-dinosaur-helped-us-prove-velociraptors-had-feathers-44788.
32 Lü, J., &Brusatte, S. L. (2015). A Large, Short-Armed, Winged Dromaeosaurid (Dinosauria: Theropoda) from the Early Cretaceous of China and Its Implications for Feather Evolution. *Scientific Reports*, 5(11775), doi:10.1038/srep11775.
33 Xu, Xing, et al. (2012). A Gigantic Feathered Dinosaur from the Lower Cretaceous of China. *Nature*, 484, 92–95.
34 Rauhut, O. et al. (2012). Exceptionally Preserved Juvenile Megalosauroid Theropod Dinosaur with Filamentous Integument from the Late Jurassic of Germany. *PNAS*, 109(29):11746-11751..
35 Swisher, Carl C.; Wang, Yuan-qing; Wang, Xiao-lin; Xu, Xing; Wang, Yuan (1999).Cretaceous Age for the Feathered Dinosaurs of Liaoning, China. *Nature*, 400 (6739): 58–61. doi:10.1038/21872.

of feathers.[36] This is a big deal, because the ornithischian dinosaurs are not supposed to be closely related to birds at all!

Dinosauria

In order to have some grasp of all this, I think it's valuable to know a little about dinosaurs.

In 1888, the British paleontologist Harry Govier Seeley classified dinosaurs into two orders – Ornithischia and Saurischia – based on their hip structure. These days, the Saurischia is considered more of a general group than an order, but the classification system still holds. The herbivorous *Triceratops* and *Stegosaurus*, club-tailed *Ankylosaurus* and rock-headed *Pachycephalosaurus* had beaks and hips that resembled bird hips. These are classified in the Ornithischia, the "bird-hipped" dinosaurs. The two-legged dinosaurs and long-neck dinosaurs were placed in the Saurischia, the "reptile-hipped" clade. Just to be confusing, the Ornithischia are *not* the dinosaurs regarded as the ancestors of birds. Nope, birds are placed firmly in the Saurischia.

My favorite dinosaurs are all in the Ornithischia clade. I always wanted the *Triceratops* to beat the T-Rex, every time. I've long been a fan of the *Stegosaurus* with the plates on its back and of *Pachycephalosaurus* with its bust-em hard head. However, they're not the subject here. Right now we need to focus on the carnivores in the Theropod clade.

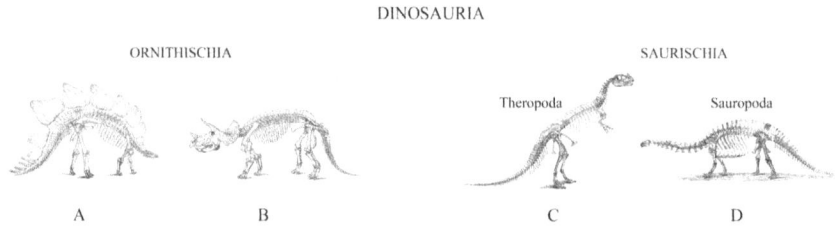

Figure 30: Reconstructions by O.C. Marsh, featuring (A) Stegosaurus ungulates *(1891) (B)* Triceratops prorsus *(1896), (C)* Ceratosaurus nasicornis *(1892), (D)* Brontosaurus excelsus *(1896) (aka* Apatosaurus excelsus*). Adapted from the Public Domain.*

---

36 Godefroit, P., et al. (2014). A Jurassic Ornithischian Dinosaur from Siberia with Both Feathers and Scales. *Science*, 345(6195): 451-455, doi:10.1126/science1253351..

The Saurischia can be subdivided into two groups: Sauropoda and Theropoda. The sauropods ("lizard footed") were the herbivorous, long-necked dinosaurs who strolled along meandering rivers on legs that resembled tree trunks. Don't we all wish we could have ridden on *Brachiosaurus* and *Diplodocus* as our gentle-giant pets?

On the other hand, the theropods ("beast footed") are the bipedal carnivores like the T-Rex and *Velociraptor*. And what did the Jurassic Park films teach us about the *Velociraptors*? We learned they are brilliant hunters that love human meat, and they fall woozy to the charms of Chris Pratt like all the other chicks.

We didn't learn that they were feathered. We didn't learn that a number of theropods had feathers of some kind or another, and even one species of ornithischian dinosaurs appears to have had feather-like structures. We didn't learn this odd fact. We just learned that humans cannot make safe dinosaur theme parks, so don't even try.

What about *Archaeopteryx*, though? Where does it fit in?

"Old Wing"

Archaeopteryx means "old wing," and there are two lines of thought on this feathered beauty. Alan Feduccia at the University of North Carolina Chapel Hill has long argued that the *Archaeopteryx* is just a perching bird with some reptilian characteristics. Feduccia - an ornithologist who specializes in bird evolution - disagrees that birds came from dinosaurs. He hates the idea that bird flight originated on the ground, or that energy-expensive feathers were originally used for mere insulation when they were so well designed for flight. In his 1996 book, *The Origin and Evolution of Birds*, Feduccia criticizes the position that birds are theropod dinosaurs.

Most creationists take Feduccia's position that the *Archaeopteryx* is a bird. Many people will say it's a bird since it has feathers and wings, and the basic definition of birds is that they have feathers. It's not even so important whether birds have hollow bones, because penguins don't have hollow bones. Birds are warm-blooded and have feathers and lay eggs. That's how we've defined birds.

Feduccia is in the minority, though, and the *Archaeopteryx* is

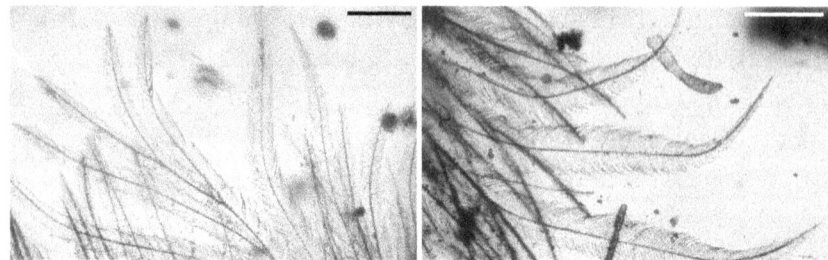

Figure 31: Photomicrographs of feathery integument in amber specimen DIP-V-15103 Left: Pale ventral feathers (white arrow indicates rachis apex). Scale bar = 1mm. Right: Dark dorsal feather. Scale bar = 0.5 mm. Adapted from Xing, L. et al. (see Footnote 38).

widely regarded as an evolutionary link between birds and dinosaurs. Hunting down its true place in biological history has proved tricky.

Mind you, many dinosaurs declared "feathered" just have a downy fuzz on their bodies, and Feduccia poo-pooed these, discounting "dinofuzz" as a simple filamentous material. A whole bunch of dinosaurs have been found with these hair-like filaments, and Feduccia rejected the tendency to put them in the feather category, telling *Discover Magazine* they looked to him like, "preserved skin fibers."[37]

However! In 2016, a bony dinosaur tail was found with delicate feathers attached, trapped and preserved in amber. It's not connected to a body, but there are enough vertebrae to categorize it as a non-avialan coelurosaur tail. The feathers were morphologically a bit different than bird feathers, but feathers none-the-less (Fig. 31).[38]

As we find more and more feathered dinosaurs, it's become apparent we can't just define a bird as a critter that has feathers. The favorite conclusion is not that dinosaurs are birds, but instead that birds are dinosaurs. We find today a common statement that birds are the final surviving branch of the dinosaur tree.

Like birds, the *Archaeopteryx* had a wishbone and long arms that could act as wings. X-ray analysis suggests the *Archaeopteryx* could

---

37 Svitil, K.A., (2011, January 31). Ornithologist and Evolutionary Biologist Alan Feduccia—Plucking Apart the Dino-Birds. *Discover*. Retrieved from https://www.discovermagazine.com/planet-earth/ornithologist-and-evolutionary-biologist-alan-feducciaplucking-apart-the-dino-birds..
38 Xing, L., et al. (2016). A Feathered Dinosaur Tail with Primitive Plumage Trapped in Mid-Cretaceous Amber. *Current Biology*, 26(24): 3352–3360, Fig. 3, doi: 10.1016/j.cub.2016.10.008.

What is the Archaeopteryx?

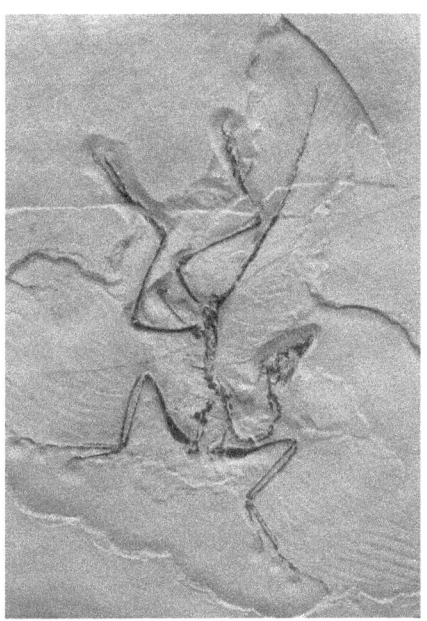
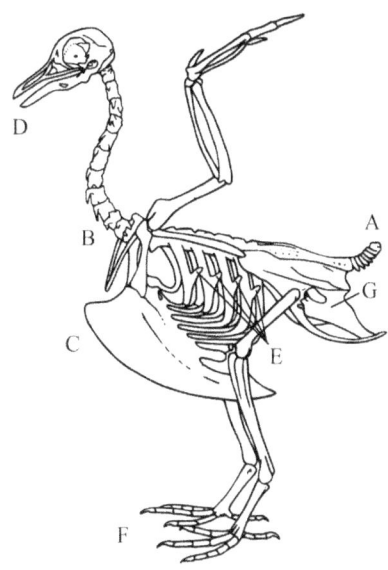

Figure 32: Left: Cast of the Berlin Specimen of Archaeopteryx lithographica. *Photo by Paul Stillwell. Right: A representation of a pigeon skeleton demonstrating typical bird anatomy, including (A) pygostyle; (B) furcula (wishbone); (C) keel; (D) beak; (E) uncinate processes on the ribs; (F) bird digits; (G) reversed pubis. Adapted from BIODIDAC.*

fly in bursts - much like a pheasant.[39] The Berlin Specimen of the *Archaeopteryx* is so well preserved that it shows the animal had body plumage and "trouser" feathers on its legs (Fig. 32).

Otherwise, the *Archaeopteryx* was nothing like today's birds. Contrast it with the skeleton of a pigeon above. Like reptiles, the *Archaeopteryx* had teeth, not a beak. It had claws at the ends of its forelimbs, a flat sternum instead of a keel, a long tail with distinct caudal vertebrae - and not the nubby pygostyle of a bird. The *Archaeopteryx* had gastralia (belly ribs) which have the distinction of not being attached to the backbone. They're "floating" bones that protect the belly. The *Archaeopteryx* ribs also don't have the hook-like uncinate processes typical of bird ribs. The *Archaeopteryx* had dinosaur-like feet, and its second toe could hyperextend to kill prey – like those vicious little *Velociraptors* in our favorite movies.

---

39 Voeten, D. F. A. E., et al. (2018). Wing Bone Geometry Reveals Active Flight in *Archaeopteryx*. *Nature Communications*, 9:923, doi:10.1038/s41467-018-03296-8.

In many ways, the *Archaeopteryx* was similar to bigger theropod dinosaurs like *Allosaurus*, which also had a wishbone. *Allosaurus* and T-Rex had belly ribs; bipedal dinosaurs are nearly the only creatures that do. T-Rex had a wishbone, but its arms – feathered or not – were clearly not for flying.

In other words, the *Archaeopteryx* had wings and feathers, but basically everything else about it was reptilian like other dinosaurs.

Alan Feduccia goes into detail on the bird-likeness of the *Archaeopteryx*, but his focus is on its feathery wings and presumed perching claws.[40] He takes issue with the ground-up model of bird flight evolution and argues that the *Archaeopteryx* was already in the trees, flying like a true bird. Analysis of the Thermopolis specimen of the *Archaeopteryx*, however, shows that its hallux stuck out to the side, like our thumb, rather than backward like the back toe of a perching bird.[41] What's more, several *Archaeopteryx* skull characteristics are like that of a theropod dinosaur.[42]

Which raises the question: do feathers on dinosaurs indicate that they are related to birds (which is the general consensus among paleontologists these days), or is this a case of "convergent evolution" in which the same characteristic appears on critters that are not closely related?

Let us keep exploring this.

Who's Your Mommy?

A 2014 article in the journal *EvoDevo*[43] offers an evolutionary tree of birds from dinos. It specifically considers the tails of ancient skeletons in the big, umbrella Paraves clade of dinosaurs. The *Velociraptor* had feathers, but it couldn't fly. *Archaeopteryx* had feathers and could fly, but it still had a long, bony tail. *Sapeornis* had flying feathers and a short tail, but it still had a lot of reptilian characteristics. *Confuciusornis* (which coexisted with *Sapeornis*) had

---

40 Feduccia, A. (1999). The Origin and Evolution of Birds. New Haven: Yale University Press.
41 Mayr, G., Pohl, B., & Peters, D.S. (2005). A Well-Preserved Archaeopteryx Specimen with Theropod Features. Science 310(5753):1483-1486, doi: 10.1126/science.1120331.
42 Rauhut, O. (2014). New Observations on the Skull of *Archaeopteryx*. Paläontologische Zeitschrift. 88:211-221, doi: 10.1007/s12542-013-0186-0.
43 Rashid, D. et al. (2014). From Dinosaurs to Birds: A Tail of Evolution. EvoDevo, 5:1-20, doi: 10.1186/2041-9139-5-25.

all kinds of bird characteristics - all of a sudden out of the blue! It looks like a kinda-sorta series of transitions trending toward modern birds.

But this kind of sequencing doesn't mean anything real. What matters is whether there's a genetic relationship between any of these guys. The fossil record is replete with convergent evolution - creatures that aren't related, but have the same body parts anyway. Modern birds that look and behave in similar ways can have significant genetic differences, so who cares if *Sapeornis* had a short tail? Especially since it co-existed with *Confuciusornis*, that Asian cutie.

Back in 1998, analysis of the mitochondrial DNA of birds showed they had multiple origins.[44] (Humans don't have that problem; our mtDNA shows that we all had a common mother "Mitochondrial Eve" way back in history.[45]) Modern gene-sequencing technology now makes whole-genome analysis possible, and a 2014 paper reported the results of a huge collaborative effort to compare the genes of 48 bird species from all manner of bird branches. To their puzzlement, the researchers found that having the same body parts and behaviors didn't necessarily mean two sets of birds had a close evolutionary relationship.[46] One of the paper's authors, Simon Ho of the University of Sydney commented:

> The team was able to work out the relationships among the major groups of modern birds, showing that our previous understanding of birds had been clouded by the appearance of similar traits and habits in distantly related groups. So while grebes and cormorants are both waterbirds with webbed feet that dive to catch their prey they are, despite these similarities, from completely distinct lineages.[47]

---

44 Mindell, D., Sorenson, M., & Dimcheff, D (1998). Multiple Independent Origins of Mitochondrial Gene Order in Birds, *PNAS*, 95(18): 10693-10697.
45 Cann, R., Stoneking, M. & Wilson, A. (1987). Mitochondrial DNA and Human Evolution. *Nature* 325:31–36, doi:10.1038/325031a0.
46 Jarvis, E. D., Mirarab, S., Aberer, A. J., Li, B., Houde, P., Li, C., ... Zhang, G. (2014). Whole-genome Analyses Resolve Early Branches in the Tree of Life of Modern Birds. *Science*, 346(6215): 1320–1331, doi:10.1126/science.1253451.
47 University of Sydney (2014, December 11). Rapid Bird Evolution after the Age of Dinosaurs Unprecedented, Study Confirms. *AAAS EurekAlert*. Retrieved from https://www.eurekalert.org/pub_releases/2014-12/uos-rbe121114.php.

These researchers worked hard to find the best way to map out the relationships of all the world's birds, and the 2014 study concludes that an explosive radiation of birds took place in a short amount of geologic time after the dinosaurs died out. More on this in a moment.

Confucius Bird

By the way, we do find birds living alongside dinosaurs in the lithified mud of the fossil record, which makes sense in light of those Cretaceous bird tracks Joe photographed (Ch. 9). For one, *Confuciusornis sanctus* dates to the Lower Cretaceous, and it looks like a bird we might see sitting in a tree today, with a beak and no teeth.[48] Hundreds of these specimens have been found near each other on bedding planes, as though the whole colony or flock was buried at one time. Unlike the *Archaeopteryx*, the *Confuciusornis* had a pubis that pointed backward like in modern birds. It had bird toes, good for grasping branches, and it would have looked like a perching bird about the size of a pigeon or small crow. It had long, asymmetric feathers and long wings. Like modern birds, it had a pygostyle, sometimes with long tail feathers, and a wide sternum with a low keel. The *Confuciusornis* did have claws at the ends of its wings as well as gastralia, which modern birds do not have, but specimens of *Confuciusornis* have also been found with uncinate processes on their ribs.[49] Nearly all modern birds have uncinate processes, and it's a distinctive trait that *Archaeopteryx* lacks. The Confucius bird is dated to just a few million years

Figure 33: Representation of plumage in Confuciusornis. Art by Velizar Simeonovski. International license CC BY 4.0.

---

48 Hou, L., Zhou, Z. Martin, L.D. and Feduccia, A. (1995). A Beaked Bird from the Jurassic of China. *Nature*, 377:616-618.
49 Chiappe, L., Ji Shu'an, Ji Qiang, J., Norell, M.A. (1999). Anatomy and Systematics of the Confuciusornithidae (Theropoda: Aves) from the Late Mesozoic of Northeastern China. *Bulletin Of The American Museum Of Natural History*, 242:32.

What is the Archaeopteryx?

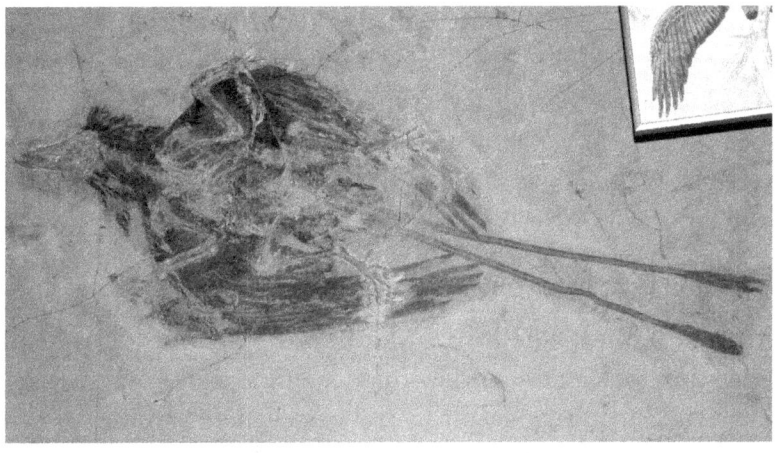

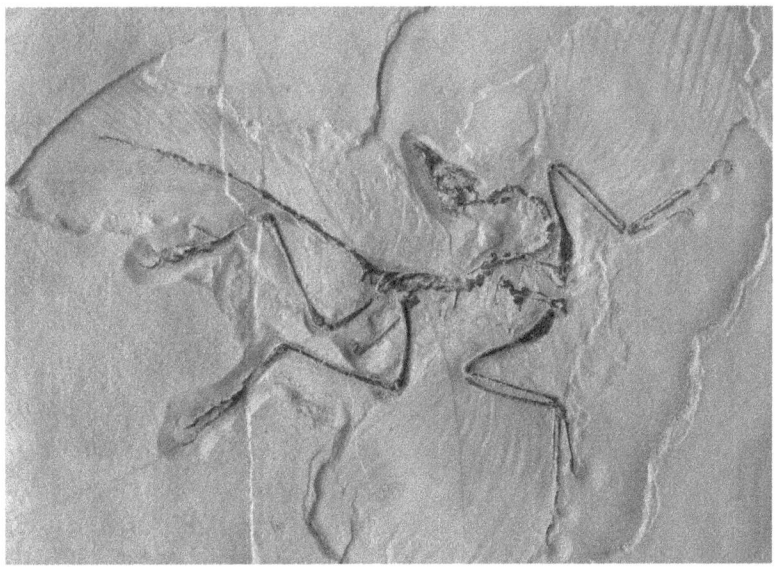

Figure 34: Top: Confuciusornis sanctus, *The Natural History Museum, Vienna. Photo by "The Paleobear" retrieved from https://www.flickr.com/photos/paleo_bear.* Bottom: Archaeopteryx lithographica, *the "Berlin specimen" cast by the author.*

after our Old Wing friend *Archaeopteryx*. In Figure 34 above, *Confuciusornis* (top) resembles a modern bird, unlike the lizard with wings (bottom). There are others. The confuciusornithid genus *Changchengornis* on display at the Geological Museum of China looks like a fluffy bird that got smushed, with limb-grasping claws and a pygostyle.[50] The experts have said that the confuciusornithids are "clearly not ornithurine birds" and have placed them in their own family within the suborder Theropoda.[51] Yet, gastralia and claws on their wings seem small issues for a bird found buried in Lower Cretaceous sediments, especially considering the existence of living birds like the hoatzin.

HOATZIN AND PLATYPUS

The "stink bird," the hoatzin, is a bizarre living bird of South America that looks like a turkey mated with a punk rocker. It isn't closely related to any other birds, and even genetics studies place it in a sister group of its very own.[52] The stink birds build their nests above water, but that's okay because baby hoatzins can swim just fine. They are also born with claws at the ends of their wings, which comes in handy when climbing back up the tree to their home nest. They grow up to be clumsy fliers at best, and unlike most other birds, they eat only leaves. Rather than breaking down food in a gizzard like most birds, they ruminate the vegetable matter like a cow. This odd bird has feathers like all other birds, but in many ways it's out there in space by itself.

Figure 35: Hoatzin from Peru. Photo by Francesco Veronesi. Reproduced under International license CC BY SA 2.0.

The hoatzin reminds me of the platypus and echidna, alone in an order of mammals called monotremes, which lay eggs like a reptile. The platypus has a flat tail like a beaver and a leathery duck-like bill with super-sense powers. It closes its eyes and ears underwater and senses its prey with the multitude of sensors in its bill!

50   Ji Qiang, Chiappe, L., & Ji Shu'an.(1999). A New Late Mesozoic Confuciusornithid Bird from China. *Journal of Vertebrate Paleontology*, 19(1): 1-7.
51   Chiapee, L., et al. (1999).
52   Prum, R.O, et al. (2015). A Comprehensive Phylogeny of Birds (Aves) Using Targeted Next-Generation DNA Sequencing. *Nature*, 526:569-573, doi:10.1038/nature15697.

In its hind legs, the male platypus possesses spurs that secrete an exceptionally painful venom. Like all mammals, it has hair and produces milk for its babies, but it has "poor" representation in the fossil record, and even suspected ancient ancestors are speculations at best.[53] It's a weird hodgepodge animal, a mixture of bits and pieces that appeared out of nowhere in the fossil record. It's a mammal! It has hair and lactates! Otherwise, it's out there in space by itself.

Is the *Archaeopteryx* a dinosaur that was a side-branch in the general trend of dinosaurs turning into birds, or is it just a feathered dinosaur? Were feathered dinosaurs related to birds at all? Or were dinosaurs like the platypus with a hodgepodge of characteristics, some shared by birds and some by reptiles?

It might be that birds are indeed the last existing branch of the dinosaur kind. That might be the case. On the other hand, the evolutionary worldview of most paleontologists might be forcing the dinosaurs into family tree connections that aren't real. Hopefully these kinds of questions can be answered as we find more and more feathered dinosaurs.

OF CABBAGES AND KINGS

We have a community of biologists and paleontologists who take for granted that all living things on earth had a common ancestor. It's assumed that lions and dogs and fish and amphibians and mosquitos all evolved from common ancestors, which ultimately evolved from good old LUCA, the "last universal common ancestor." These biologists and paleontologists are perpetually trying to draw out the family trees that connect all animals together with LUCA down at the bottom of the trunk. Because they *assume* a family tree at the get-go, they are constantly trying to map out those relationships, and it often gets convoluted.

My prediction! I predict that their efforts will work when there are real relationships involved, and their efforts to make connections will get highly problematic when those relationships cease to exist.

---

53 Rowe, T. et al (2008). The Oldest Platypus and Its Bearing on Divergence Timing of the Platypus and Echidna Clades. *PNAS*, 105(4):1238-1242, https://doi.org/10.1073/pnas.0706385105.

The researchers who labored to map out bird lineages ran into some stumbling blocks before publishing their paper in 2014, which is why bird evolution was deemed "rapid radiation" or an "explosion." I suspect that some legitimate genetic relationships were defined, and (for instance) hummingbirds might really be closely related to swifts. They might truly have had a common ancestor. However, all those bird branches were supposed to connect at the bottom somewhere, and that's where things got trippy. If evolution had taken place from a single common ancestor over a long period of time, with lots of intermediate tree branches, it should have been fairly straight-forward to demonstrate those relationships. We would have expected the bottoms of the branches to meet together with reasonable gaps due to extinctions. Instead, the various bird lineages appear to have sprung up out of nowhere.

It's like... let's say we tried to arrange our Valentine's candy into similar piles. First we tried to separate them by color, but when we put all the red candy together, we found that they were all different shapes.

So, we tried to put all the red hearts into a pile and all the red diamonds into a different pile, but when we did, we found that some in each pile were chocolate inside and some were gummies and some were hard sugar candy.

And when we opened some of the pink triangles and the yellow hearts, they were also gummies or chocolate or sugar candy inside.

In other words, the bird genomes didn't group up neatly, and convergent evolution was demanded quite a bit. The researchers were expecting a tree-like pattern, all the way to the base. It took them *years* working with big, brutal supercomputers to crunch all the data and come up with their best efforts at final results, but the results were still confusing. One of the study's authors wrote:

> ...[T]here is increasing evidence that loci can have conflicting evolutionary histories (so that their phylogenetic trees are topologically different) because of many biological causes...[54]

---

54 Mirarab, S. et al.(2014) Statistical Binning Enables an Accurate Coalescent-Based Estimation of the Avian Tree. *Science* 346(6215):1250463. doi:10.1126/science.1250463.

What were some of the causes? First, the researchers concluded that evolution happened quickly there for awhile. That is, the DNA changed in big ways. Second, some genes had a lot of variation - multiple alleles - and some of those gene variations got lost in subpopulations during the rapid radiation, something called incomplete lineage sorting (ILS). No more wing claws.

I genuinely applaud the work of the researchers and their efforts to come up with the statistically most-likely scenarios for bird evolution. I applaud their hard work, and I'm glad they did it! Somebody needed to do it! But, I'm going to be terribly gauche and suggest that they faced those contradictions and frustrations because certain groups of birds aren't related to each other. Rapid radiation is incorrect. They had no common ancestor, ever.

That's an alternate explanation that I think should be considered, and I suspect we'll find the same is true of other lineages of critters. When there are real relationships, we should be able to map them out in a fairly straight-forward manner. Of course there will be gaps or unexpected gains and losses, but following the dots will lead to X on the treasure map. When there aren't real relationships, the expected (but incorrect) genetic connections will get convoluted and deeply aggravating. That's my simple prediction, that we'll find the same problem as we go backward in the lineages of reptiles and fish and perissodactyls: there will be "rapid radiation" and severely unresolveable evolutionary histories, because certain branches never shared an ancestor.

There's an assumption in the biological community that common physical traits equal evolutionary connections, and I'm not ready to make that assumption, because of complex issues with DNA programming.

The biological community takes for granted that all animals with a backbone evolved from a common ancestor with a backbone. We have five groups of vertebrates: mammals, fish, amphibians, reptiles, and birds. We group vertebrates together because they all have a backbone. Therefore, mammals, fish, amphibians, reptiles, and birds must be physically related; they had the same great great grandma at one point – going back many hundreds of millions of years. That's the position all the respected biologists take.

The alternate view - the not-respected view - is that mammals, fish, amphibians, reptiles and birds were all created separately in the first place, and God gave them backbones and skeletal systems, because He wanted all of them to have structures that worked for the lives they needed to live. In this case, the commonality of having a backbone means very little as far as family history is concerned.

Either all vertebrates have faces with two eyes, a nose, and a mouth because they are genetically related way back in time, or they all have faces because God likes faces. I think those are our choices, and I'm willing to conclude that God likes faces. Even fruit flies and jumping spiders and zebrafish larvae have faces - due to unique coding in their hox genes, coding different than the coding that gives other creatures faces.

"Don't eat anything with a face," Dr. Stillwell once quoted his wife the vegetarian. If Genesis is history, Adam and Eve ate plants; they didn't eat things with faces.

We can't put God in a test tube, but we can see which model the evidence best fits.

1. Does the vast breadth of paleontological and genetic evidence fit Model #1 - that all life on earth sprouted and branched up from LUCA, a single source? Do all the branches connect in reasonable, traceable ways, time gaps notwithstanding?

OR.

2. Does the evidence fit Model #2 - that all creatures on earth branch from multiple origins? Do all species today sprout from one of a large number of original trunks?

The general Theory of Evolution declares the first to be true, and the Bible declares the second to be true. Which works best?

I have no problem at all with the idea that dromaeosaurs are all related to each other. Tigers, lions, lynxes, bobcats, and Morris the cat all had a common ancestor? Clydesdales, Appaloosas, Shetland Ponies and three-toed horses all had a common ancestor? Llamas and camels had a common ancestor? Okay! We can trace changes within genera and families and sleuth up why donkeys have 62 chromosomes and horses have 64. The big question is whether all these horses and cats and camels had a common ancestor with dromaeosaurs.

If birds were originally one branch of dinosaurs, and all the other dinosaurs died out, then that's fine with me. I have no problem with whatever is true. But, maybe feathers are like backbones. Maybe God created dinosaurs with feathers, and He created birds with feathers, and they aren't related. Or. Maybe God invented all sorts of different kinds of dinosaurs, and birds are one group of dinosaurs.

As I hunt through these matters, I face a serious problem. I need to discern between legitimate factual evidence on one hand and gross evolutionary assumptions on the other. The biologists and paleontologists analyzing skeletons in the literature *assume* that birds had to evolve from somewhere, and dinosaurs are the best option, so dinosaurs it is. They assume that *Archaeopteryx* had to be on its way to becoming a bird just because it had feathers and a wishbone. I'm trying to throw off all those assumptions and track down how the known facts fit together most naturally, without any shoe horns.

At this point in my research (and it might change as I collect more knowledge), it appears that the *Archaeopteryx* is a dinosaur with feathers, like the *Velociraptor* and *Deinonychus* and other dromeosaurids, and it appears that *Confuciusornis* is a full-on bird with wing claws and gastralia. And the platypus is a mammal that has venomous barbs and lays eggs, and the hoatzin is a bird with a ruminating gut.

And, for now, God only knows how any of them are related.

THE LAETOLI TRACKS

The Paluxy River dinosaur tracks are not the only possible source of controversy about the age of the earth and humanity and the meaning of life. The Laetoli tracks also offer faster heartbeats to lovers of ancient prehistory.

I decided that having this entire section in the main body of the book was a little much, so I've moved the expanded version of the chapter here to the end of the Appendix, to the land of free and the home of the brave. This is the full version of the original "Chapter 27: The Laetoli Tracks" for all those who like my daring analysis - or who want to tell me why I'm wrong. Either works for me.

Photos abound of footprints preserved in an ancient layer of volcanic ash at the Laetoli site in Tanzania. The tracks are famous (in fact, I have the November 2013 issue of *Scientific American* here beside me, and the Laetoli prints conveniently appear on page four). The three sets of footprints look like human prints, which Mary Leakey recognized when she found them in the late 1970s, saying:

> Note that the longitudinal arch of the foot is well developed and resembles that of modern man, and the great toe is parallel to the other toes.[55]

> ...[I]t is immediately evident that Pliocene hominids at Laetoli had achieved a fully upright, bipedal and free-striding gait...[56]

Most studies of the Laetoli tracks focus on the smallest set of prints, the G1 trail, which was probably made by a juvenile. Until recently it's been difficult to analyze the larger G2 and G3 tracks, because the G3 individual walked inside the footprints of the G2 leader. A recent 3D analysis of these tracks has been able to separate

---

55 Leakey, M. & Hay, R. (1979). Pliocene Footprints in the Laetolil Beds at Laetoli, Northern Tanzania. *Nature* (278): 320.
56 Ibid., 323.

the prints, giving us greater insight into the adult foot morphology as well, which is great.[57]

Russell Tuttle, anthropologist at the University of Chicago and longtime editor of the *International Journal of Primatology* carefully examined the prints and concluded in 1990:

> In sum, the 3.5-million-year-old footprint traits at Laetoli site G resemble those of habitually unshod modern humans. None of their features suggest that the Laetoli hominids were less capable bipeds than we are. If the G footprints were not known to be so old, we would readily conclude that they had been made by a member of our genus, *Homo*...[58]

Tuttle thought researchers were hasty to associate the footprints with the australopithecines, but he also didn't think the prints were made by modern humans. He later said, "What that creature was above the ankles, God knows."[59]

That's just the problem. The prints are dated to 3.5 million years ago and so are attributed to an ancient "hominid" - a group that includes both humans and apes. The prints aren't found in dinosaur-age layers, but they *are* found in the same layer as certain australopithecines, the current favorite evolutionary relative between apes and humans. *Homo erectus* wasn't supposed to have evolved until about 2 million years ago, but the famous *Australopithecus afarensis* specimen "Lucy" was unearthed at the Hadar site in Ethiopia in 1974 and has been dated to 3.2 million years. It's widely believed that her relatives in Tanzania made the Laetoli prints.

There has been a long-standing controversy over Lucy's walking style. Her finger and toe bones are curved like an ape's, which indicates that she used her feet for gripping, as in tree climbing. Tuttle has argued that if the Laetoli prints were made by *A. afarensis*, they would show evidence of those curved toes, and they don't.[60]

Despite Tuttle's disapproval, the footprints are commonly

---

57 Bennett, *et al*. (2016). Laetoli's Lost Tracks: 3D Generated Mean Shape and Missing Footprints. *Scientific Reports*, 6:21916; doi: 10.1038/srep21916.
58 Tuttle, R. (1990). The Pitted Pattern of Laetoli Feet. *Natural History*, 99(3): 64.
59 Kelly, J. (2014, Jan/Feb). Only Human. *The University of Chicago Magazine*.
60 Lewin, R. (1983) Were Lucy's Feet Made for Walking? *Science*, 220:700-702.

credited to Lucy's local kin, other australopithecines in the area.

Another major problem with this view is that Lucy's midfoot bones are missing, so we can't say whether she had human-like feet or ape-like feet. (If you go into any natural history museum and see a model of Lucy, they always give her human feet, but those are shameless assumptions at work.)

Be aware, paleoanthropologists don't believe we evolved from apes - or even straight from australopithecines. They believe we had a common ancestor with apes, and then farther along in history we had a common ancestor with the likes of Lucy. They believe that Lucy had a variety of qualities that made her more human-like, primarily the possibility that she walked upright on two feet. She was still small and had long arms and a small brain and otherwise acted like a chimpanzee.

Apes' feet are quite a bit different than human feet. Their big toe is splayed out like the thumb on a hand. Apes' feet are flexible for gripping tree trunks and branches, and they have no arch. The Laetoli footprints are like ours, with arches and forward-pointing toes. The hallux (big toe) on both G1 and G3 is somewhat "adducted" - that is, it points outward - but it is still human-like and not ape-like. G3 looks nearly modern.[61] The spacing of the footprints demonstrates a human-like gait, and there are no knuckle marks or other ape-like signs on the trail.

In March, 2010, David A. Raichlen and his team at the University of Arizona published a paper that compared the Laetoli prints to those of modern humans. Raichlen has published extensively on bipedal locomotion, and his team concluded that the Laetoli prints represent a true human-like gait and not the bent-knee bent-hip movement of apes:

> The relative toe depths of the Laetoli prints show that, by 3.6 Ma, fully extended limb bipedal gait had evolved. Thus, our results provide the earliest unequivocal evidence of human-like bipedalism in the fossil record...

---

61 Bennett *et al.* 2016

The Laetoli Tracks

...By 3.6 Ma, hominins at Laetoli, Tanzania walked with modern human-like hind limb biomechanics...[62]

If australopithecines made the Laetoli prints, they had to have had human-like feet and a human-like gait. But, did they?

LUCY'S FEET

I've been reading through the literature, trying to find actual fossil evidence that *A. afarensis* feet had forward-pointing toes and an arch. There is a foot specimen AL 333-115 that includes the phalanges and metatarsal heads (the toes) of an australopithecine, but these don't fit the Laetoli prints because they are long and curved like an ape's foot.[63] The best I've found is a fourth metatarsal (foot) bone sifted from tailings at the Hadar site. It wasn't discovered connected with any particular skeleton; it's just a single foot bone found among hundreds of other bits and pieces.

Ward *et al* note that "Skeletal evidence for the presence of pedal arches in *A. afarensis* has been ambiguous, because key bones from the midfoot have been lacking."[64] That makes this single foot bone highly significant. Ward's team carefully compared this new bone to the fourth metatarsal of chimps, gorillas and humans and concluded that it was more human-like than ape-like. In fact, the measurements show that it is just like a human bone.

Unfortunately, we can't be certain which creature actually produced the bone. It was discovered, "during sieving of eroded Denen Dora 2 submember surface deposits of the Hadar Formation."[65] Hundreds of australopithecine bones and multitudes of various other animal specimens have been found at the Hadar site. Unless the bone was found articulated (connected) as part of an australopithecine skeleton, the assumption should be that a human bone came from a human.

---

62 Raichlen DA, et al. (2010). Laetoli Footprints Preserve Earliest Direct Evidence of Human-Like Bipedal Biomechanics. *PLoS ONE*, 5(3): e9769.
63 Susman R.L., Stern, Jr. J.T., and Jungers W.L. (1984). Arboreality and Bipedality in the Hadar Hominids. *Folia Primatol*, 43:113-156.
64 Ward, C.V. et al. (2011). Complete Fourth Metatarsal and Arches in the Foot of *Australopithecus afarensis*. *Science*, 331: 750-753.
65 Ibid., 750.

And! Another *Homo* (human) fossil has also been found at that locality, along with rudimentary stone tools. In 1996, a maxilla (upper jaw) dated to 2.33 million years and assigned to the *Homo* genus was found tumbled into a gully from the upper Kadar Hadar Member (which sits above the Denen Dora Member in certain places).[66] Which means other *Homo* bones could be in the mix of stuff that has eroded from surface deposits over the ages, and the researchers have to be careful about their assumptions. A human-like foot bone suggests humans lived in the area. It does *not* suggest that australopithecines had human-like feet when everything else we know about their feet indicates they didn't.

What information do we have about Lucy's feet? In 2009, a group of researchers found a fourth metatarsal bone that dates to the time of Lucy, but it's got characteristics that indicate a foot with an opposable big toe, good for gripping trees like an ape foot.[67] It was therefore *not* assigned to *A. afarensis*, but to another undesignated hominin living in the area at the same time. A description of the bone was published in *Nature* in March 2012, and in an announcement by Johns Hopkins, co-author Naomi Levin states:

> The foot belonged to a hominin species – not yet named – that overlaps in age with Lucy... Although it was found in a neighboring project area that is relatively close to the Lucy fossil site, it does not look like an *A. afarensis* foot."[68]

Woah woah woah. What she should rather say is, "it does not look like what we expected an *A. afarensis* foot to look like." This sort of thing concerns me. Don't they have a solid, objective methodology in place for assigning bones to different species?

Heels and ankles and lower leg bones have been found in the Hadar Formation as well, and analyses have been done on them. Bruce Latimer and Owen Lovejoy have produced a series of articles

---

66 Kimbel, W. (1996). Late Pliocene Homo and Oldowan Tools from the Hadar Formation (Kada Hadar Member), Ethiopia. *Journal of Human Evolution*, 31: 549-561.
67 Yohannes Haile-Selassie, et al. (2012). A New Hominin Foot from Ethiopia Shows Multiple Pliocene Bipedal Adaptations. *Nature*, 483:565-569.
68 De Nike, L. (2012, March 30). Newly Discovered Foot Points to a New Kid on the Hominin Block. Retrieved December 27, 2014, from http://releases.jhu.edu/2012/03/30/newly-discovered-foot-points-to-a-new-kid-on-the-hominin-block/

on the *A. afarensis* lower-limb structure, and these experts are huge fans of Lucy as a human ancestor who walked on two legs.[69]

Not everybody agrees with Latimer and Lovejoy, though. Anatomist William Jungers long ago noted that Lucy and her kin may have walked upright, but they didn't walk like humans. He wrote in a letter to *Nature*:

> The bodily proportions of Lucy are not incompatible with some form of bipedal locomotion, but kinematic identity and functional equivalence with the bipedal gait of modern humans seem highly improbable.[70]

It's clear from Lucy's hip structure that she did not have a human's free-swinging gait. She waddled when she walked.

A Class of Its Own

Australopithecines were being studied long before Lucy. Anatomist E.H. Ashton studied primate anatomy since the 1950s with fellow experts like Charles Oxnard and Solly Lord Zuckerman. These were famous zoologists. They carefully worked to compare australopithecines to both humans and apes. Ashton took a very systematic, measurements-based approach to analyzing the anatomy of these creatures. That is, he measured the australopithecine bones and compared them to the normal ranges for human bones versus the ranges for ape bones.

When he did this, Ashton determined that the australopithecines were very like apes in some respects and very like humans in other respects, but they were so different from both apes and humans that they should not be classified in the human line at all.[71] Ashton ends this particular article by saying:

---

69 e.g. Latimer, B., and Lovejoy, C. (1990). The Hallucial Tarso-Metatarsal Joint in *Australopithecus afarensis*. *American. Journal of Physical Anthropology*, 82(2): 125-133. Or Latimer, B., and Lovejoy, C. (1989). The Calcaneus of *Australopithecus afarensis* and Its Implications for the Evolution of Bipedalism. *American Journal of Physical Anthropology*, 78(3): 369-386.
70 Jungers, W. (1982). Lucy's Limbs: Skeletal Allometry and Locomotion in *Australopithecus afarensis*. *Nature*, 297: 676-678.
71 Ashton, E.H (1981). Primate Locomotion: Some Problems in Analysis and Interpretation. *Philosophical Transactions of the Royal Society B*, 292: 77-87.

A detailed and highly critical appraisal of the available osteological evidence has shown, once again, many functionally significant contrasts with man, together with a total functional complex that suggests 'usage as in an arboreal species that also walks bipedally with flattened arches (like a chimpanzee or gorilla) rather than with the high arches of Man.'

Such a creature would have been very different from all living primates (human and subhuman). Whether or not its gait could have been ancestral to the human type of bipedalism remains indeterminate.[72]

Lucy might have walked upright after a fashion, but it's become pretty clear that her kind was partially arboreal and climbed around in trees as well.[73] It would make sense for her to still have flat feet, because flexible flat feet are useful for tree climbing. The experts discuss whether Lucy had arches or flat feet because they are missing the bones that would give them a definitive answer. Happily, other nearly complete australopithecine specimens have been discovered since Lucy.

Little Foot

Stw 573 "Little Foot" is a fantastic little australopithecine specimen found buried in sediments in the Sterkfontein caves of Gauteng, South Africa. She's been dated to 3.6 million years,[74] and she's one of the most complete australopithecines ever found, with a collection of extremely important parts, including skull, ribs, pelvic pieces, a complete articulated forearm, and leg and feet bones! At first, lead excavator Ron Clarke argued that Little Foot's feet could have made the Laetoli prints. However, as additional foot bones were examined, the more ape-like her features appeared. Little Foot had

---

72   Ibid., 85-86.
73   D. J. Green, Z. Alemseged (2012). *Australopithecus afarensis* Scapular Ontogeny, Function, and the Role of Climbing in Human Evolution. *Science*, 338 (6106): 514.
74   Darryl E Granger et al. (June 4, 2015) New cosmogenic burial ages for Sterkfontein Member 2 Australopithecus and Member 5 Oldowan. *Nature*, 522: 85-88.

an opposable big toe like an ape, and she seemed well-adapted for climbing around in trees.[75]

Has Little Foot changed the prevailing theories? Nope. She was originally classified as an *A. afarensis* specimen, but close examination of the skull and differences in the cheeks, jaw, teeth and general facial shape caused Clarke to place Little Foot in a new species of the same genus: *Australopithecus prometheus*.[76] Fair enough. Yet, Little Foot is our best view into the foot structure of an australopithecine so far, and she has ape-like feet and not human feet.

This is precisely why I have trust issues. Is there actually solid evidence that Lucy's kin had anything to do with those footprints down in Laetoli? It bothers me that so many people in the literature assume that australopithecines made the Laetoli footprints, because that's the explanation that fits what they already believe. I see no evidence that Lucy had human feet. This is ridiculous. Evidence that points to the australopithecines as ape-like appears to be downgraded and evidence that suggests they were human-like appears to be severely exaggerated.

Here's a good example. Remember, the Laetoli prints show no evidence of knuckle-drag marks. Brian Richmond and David Strait wrote in *Nature* in March of 2000 that the wrists of *A. afarensis* indicated that these animals were knuckle-walkers at one point.[77] Richmond and Strait trust those who say that *A. afarensis* walked upright. Therefore, they don't suggest that *A. afarensis* was a knuckle walker itself, even though its wrists were set up for knuckle-walking. They simply argue that *A. afarensis* had knuckle-walkers in its family history.

It seems well-established that Lucy had the hip and ankle structure to walk upright (in her own way). And maybe Lucy only

---

75 Cf. Oliwenstein, L. (1995). New Foot Steps Into Walking Debate. *Science*, 269(5223):476-477 and Protsch von Zieten, R. & Clarke, R.J. (2003). The Oldest Complete Skeleton of an "*Australopithecus*" in Africa (StW 573). *Anthropologischer Anzeiger*, 61(1):7-17.
76 Clarke, R.J. (1998). First Ever Discovery of a Well-Preserved Skull and Associated Skeleton of an *Australopithecus*. *South African Journal of Science*, 94: 460-463. And Clarke, R.J., (2008). Latest Information on Sterkfontein's *Australopithecus* Skeleton and a New Look at *Australopithecus*. *South African Journal of Science*, 104(11-12): 443-449.
77 Richmond, B., & Strait, D. (2000). Evidence That Humans Evolved from a Knuckle-Walking Ancestor. *Nature*, 404: 382-385.

walked upright. But, maybe she did a bit of both. After all, she had the hand structure for knuckle-walking just as much as she had the leg structure that allowed her to walk upright (in her own way). We could as easily argue that Lucy loped around on her knuckles, but her ancestors once walked upright.

I know I'm going on and on, but this whole thing seems so crazy to me. It gets even worse when we compare the way different skeletal finds are analyzed. For instance, a tall specimen called Big Man has been touted as an australopithecine who walked and ran upright like a regular human. Yet, from what I can tell, Big Man appears to be a human found in an unexpected layer.

THE BIG GUY

Big Man's official name is KSD-VP-1/1. He was found in Ethiopia in February of 2005 and described in *PNAS* in 2010.[78] Because of his large size, this partial skeleton has been nicknamed *Kadanuumuu* - "Big Man" in Ethiopia's Afar language. Like Little Foot, Big Man has been age-dated to about 3.6 million years, close to the same age as the Laetoli tracks.

Big Man is completely headless, but he's been called an *A. afarensis* specimen because he possesses some pelvic attributes like Lucy. He is otherwise quite a lot different than Lucy, which the authors explain as a result of his being both male and much larger. Yet, whereas Lucy was beginning to look a whole lot like a chimpanzee, Big Man stands up as the sort of person who could make the Laetoli prints. (As a woman, I'm insulted.)

That's just the problem, though. Big Man looks so much like a man, it's reasonable to suggest that he was actually a man. The authors readily admit his pelvic characteristics are much like the *Homo erectus* specimen BSN49/P27a–d.[79] His ribs are in the human range. His scapula is human-like and his long leg falls into a human range. The authors willingly show how much Big Man is like a man, even noting details like, "The ulnar tuberosity is well preserved and

---

78 Haile-Selassie, Y. et al. (2010). An Early *Australopithecus afarensis* Postcranium from Woranso-Mille, Ethiopia. PNAS, 107(27):12121-12126.
79 Ibid., 12126.

is most similar to those of humans."[80]

With all their data, the authors conclude that Big Man demonstrates how human-like the australopithecines were. They don't for a moment suppose that they simply uncovered some *Homo erectus* fellow buried in an unexpected layer.

Famous anthropologist Zeresenay Alemseged doesn't suggest that Big Man was actually a human, but he does question Kadanuumuu's identity. He told *National Geographic*:

> With all cranial and dental elements missing, there is no compelling evidence to attribute it *A. afarensis* and not to other known species from around the same time, including *Kenyanthropus platyops* and *A. anamensis*...[81]

Alemseged is the anthropologist at the California Academy of Sciences in San Francisco who found the famous *A. afarensis* baby Selam (which, unlike Big Man, has gorilla-like shoulder blades). Alemseged has the right to offer input. But, of course he's not suggesting the alternative that I think we should consider, that Kadanuumuu was a human who went and died in the "wrong" sediment layer.

The PBS website says something telling about the Laetoli prints in an educational page made for young people:

> The footprints also look remarkably like a human's. In fact, they looked so human-like, some scientists had a hard time believing that they were made by *Australopithecus afarensis* (Lucy's species), the only human ancestor known to have lived at the time."[82]

Nobody says the simplest, most intuitive alternative: "Maybe we've overlooked something fundamental, and we were wrong about that geologic layer. Maybe humans walked across the ash layer in

---

80 Ibid., 12124.
81 Than, K. (2010, June 21). "Lucy" Kin Pushes Back Evolution of Upright Walking? *National Geographic News*. Last accessed Jan 4, 2015 at http://news.nationalgeographic.com/news/2010/06/100621-lucy-early-humans-walking-upright-science/
82 "A Science Odyssey: You Try It: Human Evolution: Fossil" - PBS http://www.pbs.org/wgbh/aso/tryit/evolution/footprints.html, last accessed February 21, 2014.

that location at that time."

I get it. There are all kinds of pieces of geologic information battering at each other here, but that doesn't mean we should shoehorn data where they don't really fit. Sometimes we have to leave holes in our jigsaw puzzle.

I obviously favor the dreadfully unpopular position that somebody in the *Homo* genus made the Laetoli prints. Obviously. But, I favor that position because I think it's honestly the most reasonable place to start. We know that humans have human feet, but it's a vast stretch to say Lucy had human feet. I think it's wrong to dismiss Kadanuumuu completely from the genus *Homo* because he's too "old." I feel like I'm dealing with the Emperor's New Clothes here. It's reasonable to suggest that a human had human ribs, shoulders and legs. The only downfall is that the "humans made these human prints" explanation doesn't fit the popular view of human origins, and scientists don't change their models easily.

I'm not in any position to analyze all the data on my own. I'm only one person, and I just can't do it. I need to be able to trust my fellow scientists, but when it comes to the issue of the history of humanity on the earth, speculation seems horribly tangled up with obvious preconceptions. It's nearly impossible to trust anybody, I don't care who they are. These issues are huge and pluck at the very Achilles tendons of the religious and philosophical feet we use to walk.

Yet, all the facts can line up if we get enough of them. If we had all the pieces, they would fit together correctly. What does that picture look like? What is the truth about those tracks along the Paluxy River? What's the truth about the tracks in the Laetoli? Who really trespassed through those places, making things confusing for everybody?

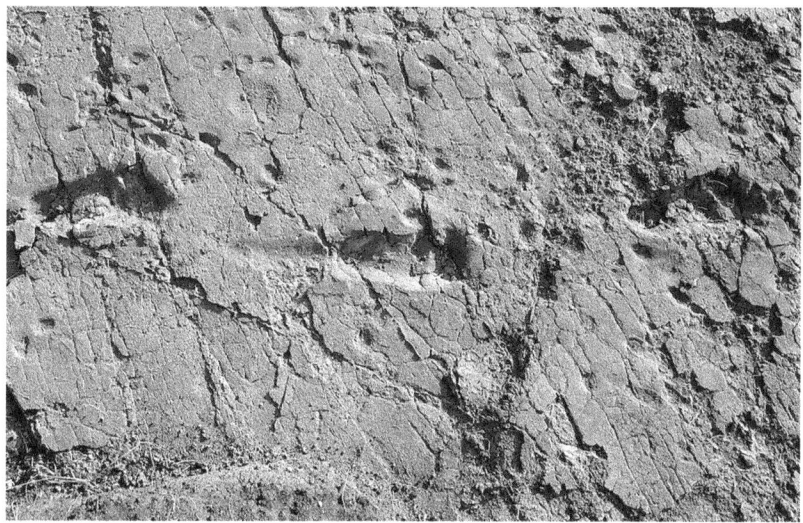

Figure 36: Southern part of hominin trackway in Laetoli site test-pit L8. Cropped by author. Credit: Masao, F. et al.(2016) New footprints from Laetoli (Tanzania) provide evidence for marked body size variation in early hominins eLife 5:e19568 https://doi.org/10.7554/eLife.19568. Licensed under Creative Commons Attribution 4.0

Books in the Science & Wonders Series

Volume 1: On The Edge of the Chasm

Volume 2: The Light, The Heat

Volume 3: As X Goes to Infinity

Volumes 4 and 5 are anticipated when the full story of Dr. Stillwell comes to completion. As of this printing, the author doesn't know how it ends.

Also by Amy Joy Hess

Gun Shot Witness: The Tim Remington Story
The true story of Tim Remington, a pastor shot six times with .45 caliber hollow point bullets - by a man who believed Tim was an alien from Mars. Tim survived and returned to his church full of people God rescued from the darkest of places.

Amy Joy and her children in 2009.
She lives in the mountains of northern Idaho.

www.ingramcontent.com/pod-product-compliance
Lightning Source LLC
Chambersburg PA
CBHW070638050426
42451CB00008B/211